Create, Share, and Save Money Using Open-Source Projects

About the Author

Joshua M. Pearce is the Richard Witte Professor of Materials Science and Engineering and is cross-appointed in the Department of Electrical and Computer Engineering at the Michigan Technological University, where he directs the Michigan Tech Open Sustainability Technology (MOST) Lab (www.appropedia.org/MOST). He is a Fulbright–Aalto University  Distinguished Chair alumnus and is also currently a visiting professor of photovoltaics and nanoengineering at Aalto University in Finland, as well as a visiting professor with the Research Team on Innovative Processes (ERPI) at the University of Lorraine in France. Pearce's research concentrates on the use of open-source-appropriate technology to find collaborative solutions to problems in sustainability and poverty reduction. It also encompasses areas of electronic device physics and materials engineering of solar photovoltaic cells and RepRap 3D printing, and includes applied sustainability and energy policy. His research group is well-known for releasing innovations with open-source licenses that eviscerate the cost of science. Pearce's work is regularly covered by the international and national press and is continually ranked in the top 0.1 percent on Academia.edu (mtu.academia.edu/JoshuaPearce). He is also the faculty advisor for the Michigan Tech Open Source Hardware Enterprise. In addition, he is founding co-editor-in-chief of *HardwareX*, a journal dedicated to open-source scientific hardware, and author of *Open-Source Lab: How to Build Your Own Hardware and Reduce Research Costs* (www.appropedia.org/Open-source_Lab).

Create, Share, and Save Money Using Open-Source Projects

Joshua M. Pearce

New York   Chicago   San Francisco   Athens   London   Madrid
Mexico City   Milan   New Delhi   Singapore   Sydney   Toronto

Library of Congress Control Number: 2020943744

1 2 3 4 5 6 7 8 9   LOV   25 24 23 22 21 20

ISBN   978-1-260-46176-3
MHID   1-260-46176-9

This book is printed on acid-free paper.

Sponsoring Editor
Lara Zoble

Copy Editor
James Madru

Editorial Supervisor
Stephen M. Smith

Proofreader
Michael McGee

Production Supervisor
Lynn M. Messina

Indexer
Karin Arrigoni

Acquisitions Coordinator
Elizabeth M. Houde

Art Director, Cover
Jeff Weeks

Project Manager
Patricia Wallenburg, TypeWriting

Composition
TypeWriting

Contents

Acknowledgments

First and foremost, on a personal note, I thank my wonderful wife, Jen, for putting up with the long hours of my writing this text and all the craziness that comes with it. In addition, I also thank Jen for driving as I write this now in the car, for turning various parts of our house into mini-makerspaces to work on projects, and for her support and her reading and critiquing of this manuscript. She is my most beloved and harshest critic—if you ever think your creation is good enough, you have not met her.

I also thank my children, Emily, Jerome, Vincent, and Dominic, who actually helped me make my first 3D printer and create some of the examples in this book, as well as various "epic makes" throughout the years.

Thanks also goes to the rest of my family for their support and encouragement: Mom and Dad, as well as siblings Solomon, Mary Rachel, and Elijah, who independently invented new ways to kill room lights late at night after reading for hours and getting too tired to get out of bed. Special thanks go to Mary Rachel for giving me a copy of *Makers* by Cory Doctorow, whose fiction inspired me to try to make some of the better parts real in this book.

In addition, I thank McGraw Hill for having the foresight to publish this book, and I especially thank Lara Zoble for making the book a reality.

This book was truly a massive international and asynchronous collaboration that goes back years and contains the ingenious and incredibly useful and beautiful contributions (from art to computer code) of people with whom I have worked closely, and of many whom I have never met (or may only know of through their esoteric internet handles that you will find scattered throughout these pages).

I also thank the past and present members of my own research group, the Michigan Tech Open Sustainability Technology (MOST) Lab, for their fruitful collaboration and the fun of creating new things to help people.

For their ongoing support, I thank the Appropedia community. A giant thank you also goes to the entire GNU/Linux community for really showing us what is possible when we all work together and providing us with the free software on which we rely. In addition, I'd like to thank Arduino founders Massimo Banzi and David Cuartielles and all their collaborators for making the control of any type of equipment easy and fun. The entire world owes a great deal of thanks to Adrian Bowyer and his many collaborators for making the RepRap project into the incredible success it is. I thank all the fantastic open-source software and hardware individuals, groups, and companies that keep enabling us to reach higher; the Open Source Hardware Association and all its members; and all those who have shared their brilliance with the world and helped make so many items easy to create because of their contributions in all the varied fields discussed in this book. Special thanks also go to everyone who provided examples that are cited or shown in these pages. Finally, I thank the growing number of makers in the burgeoning maker community who inspire and teach us all.

Disclaimer

Although many brilliant sharing people from all over the world contributed to the contents of this book, all errors and omissions are mine alone. The technologies described in this book are constantly changing, and while every effort has been made to ensure accuracy, it is always best to go directly to the sources for the most up-to-date information on the various open-source hardware projects described. Wherever possible, hyperlinks are shown in the references and will be enabled on the Appropedia splash page for the book at www.appropedia.org/Create.

Finally, if any of the hardware or software is not good enough for you, remember, it's free, so quit whining and make it better!

CHAPTER 1

Introduction to the Open-Source Philosophy and the Benefits of Sharing

Prepare to explore an amazing collection of open-source projects shared by others—that literally create millions of dollars of wealth. You can do all these projects yourself by following in the footsteps of other Creative Commons users. Perhaps best of all—it is all free! Regardless of your experience, in doing your own projects, you have already benefited from this open-source sharing, even without knowing it. The internet is perhaps the best example because it is a massive collection of wildly successful open-source projects put together by millions of people sharing their time and talent. In these pages, you will learn about how to be a more proactive sharer in the Creative Commons to reap the benefits of enormous wealth for yourself and others, even if you have only a modest amount of money to start.

You can have all of this free stuff with the compliments of the global sharing community, which will be better for having you as an active member. Some of the creations are literally free, but even those that involve material supplies (such as hardware projects) will save you substantial amounts of money. There is so much free stuff that it may be tempting to grab as much free hardware and software as you can and then sit back and revel in your savings. Of course, you might never bother to share your own work. This would be a big mistake. Simply taking will unquestionably save you a lot of money, but then you will lose the opportunities

that will offer you the greatest value—when others specifically help you.

It is great to get free stuff. I love it, too. It is far better, however, when you entice the sharing community to help you directly. You do this by sharing your own work with them for free. You can think of this as preemptive "paying it forward." It is the right thing to do. This chapter discusses why you should want to be nice and share your work with the global community.

Why We Share Aggressively

Not only should you seek to share your work, you should do it aggressively, and share it as widely as possible. There are many valuable benefits to aggressively sharing your own work in the global open-source community. Let me tell you about four.

First, when you post your work freely online, there is the potential for massive peer review. *Peer review* is a fancy way of describing having others take a look at your work. This starts from perhaps simple compliments and comments or questions about your artwork. This feels nice and may be modestly helpful. It can, however, be far more useful. For example, have you ever had a cool idea for a new technology and wondered whether it could be brought to life—but you didn't have the money or time to bring in an engineering firm to have a go at it? Now you can. By sharing your idea in the appropriate circles, that idea can go all the way to detailed technical discussions of your new technology designs by trained engineers—for free! Sharing gives you an avenue to new friends who share your interests and passions. Feedback often comes from others with the same interests. These people can be amateurs or even professionals in your field of interest. Their feedback can improve your skills or even your specific projects. Often, truly different perspectives from people living all over the world can be invaluable in themselves.

Second, by sharing, you are gaining visibility for your interests and passions. Ironically, freely sharing your work also could lead to you getting a job! If you are considering working or volunteering for a business, nonprofit organization (NPO), or community organization that could use your talents and skills, you are essentially advertising your skills. The open-source community has helped my engineering students when talking with recruiters, because they are able to talk about their project in depth, and the recruiters can see for themselves what the students have worked on. For example, it is quite common now for software programmers to advertise their skills by helping on open-source software projects. This "advertising" can be useful in helping you get a job or recruit collaborators, customers, and even employees.

Third, when you provide high-quality documentation for your creations, they can also be used as educational aides. This has the benefit that young users and those training in your field of interest can learn "your way" of doing things. Thus, future collaborators or employees can be trained using your techniques. The most successful open-source projects are those that became a platform. They provide a base that others can use to build on—often in unexpected ways.

For example, one of the first designs I posted online that could be digitally manufactured by others was a humble fabric fastener. This two-piece device allows you to get a firm hold on any type of fabric or plastic sheet without puncturing it. The first part of the device, which I called the *lock* (shown in blue), fits closely into the base, which I call the *star* because of its resemblance to a ninja throwing star (shown in red; see Figure 1.1). To use the device, you wrap the lock in the fabric with its flat side touching the fabric. Then you push the lock and the fabric through the hole in the star. Finally, turn the lock piece 90 degrees, and pull it into place (as shown in the upper right in Figure 1.1, where it is grabbing a Scottish blanket). To take the fastener off, just reverse the steps.

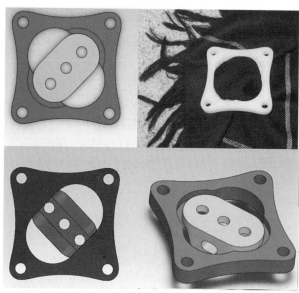

Figure 1.1   Open-source textile fastener. (CC BY-SA) www.appropedia.org/Printable_non-puncturing_textile_fastener

How does this work? Because the fabric is in contact with the assembly over a relatively large surface area, all the forces are spread out. What this means practically is that even relatively delicate fabrics will not rip unless you try something absurd. The large surface area allows the applied force in an application such as a stretcher to be spread out far enough that the fabric is prevented from ripping or sliding off. When an external force is applied (such as when you lift the stretcher), you are actually pulling the two pieces of the fastener together. This holds the fabric more tightly, rather than pulling the pieces apart. This clever little design will not accidentally fall apart unless the force is removed and considerable effort is put into unlocking the device. I found that the device could work with any type of textile from blankets and clothing material to tarps and tents. The clearance between the lock and the star will need to be adjusted if you are working with either very thin (e.g., silk) or very thick (e.g., a quilt) materials. This is easy to do in any free slicing software for 3D printers (which we will discuss in detail in Chapter 12). Both the lock and the star have multiple holes that can be used either with tie lines or nails depending on your application. This device has numerous useful applications around the home, no matter where you live. I came up with what I thought was a reasonably exhaustive list of applications:

- Stretcher: Six fasteners, a blanket, and two poles

- Hammock: Two fasteners and a blanket

- Emergency tent: Four fasteners and a plastic sheet or tarp

- Greenhouse or row guard: Use however many fasteners you need and clear plastic to create a greenhouse.

- Food transport: Use some form of fabric and four fasteners to make a flexible box for carrying such things as vegetables.

- Improved tarps:
  - You can use a fastener to fix a grommet that has popped out of a normal tarp.

  - You can also use a number of fasteners to join multiple tarps together into a bigger tarp.

  - You can create a custom waterproof cover.

- You can use the fasteners to hold the tarp in a custom way for another function, such as rainwater collection or a roof leak or a boat protector.

Oh—but boy was I wrong! By the end of the month in which I posted the design, someone had turned it into a remote control holder for the Wii to improve game play by attaching it to a glove—all without destroying the glove, so that the holder could be snapped on and off. Then Josh Smith wanted a way to turn a blanket into a cloak. He designed a blanket cloak clasp using my design and a bear sculpture from another open-source sharer to make a really cool reusable cloak clasp that looks like a bear head (see Figure 1.2). The design works nicely because you can switch in another blanket when the first one gets dirty. Or, if you want to use the cloak as a blanket again, after Halloween or your cosplay outing, just take the clasp off. This was cool—and certainly not the direction I was thinking to go with the simple design.

However, my little design was not done yet. In its small way, it helped a much more sophisticated project that is anything but a game or a fun way to keep your shoulders warm. My fabric fastener was used to improve the fasteners of the Waterpod. The Waterpod (as seen in a model in Figure 1.3) is an open-source floating structure designed as a futuristic habitat. It is an experimental platform for assessing the design and efficacy of living systems that is using shared ideas in many different ways. The main goal of the Waterpod project is to create an autonomous, fully functional marine shelter. This is a pretty ambitious project. I am happy to know that I helped it in a small way, just as I am happy to help Wii gamers improve their play or to help make cool blanket cloak holders. Helping others feels good. This is one of the beauties of the open-source approach.

Figure 1.2 Open-source bear-themed coat clasp. The design is by Josh Smith. ([CC BY] www.thingiverse.com/thing:3463125), and the image is compliments of E. Bow Pearce (CC BY).

Figure 1.3 Waterpod project concept rendering. (Antony Kim, CC BY-SA) http://www.appropedia.org/images/9/9d/Waterpod2.jpg

When you share something in the global commons, you are planting a seed. Others may help that seed grow into a mighty tree—perhaps beyond your wildest ideas. Through a modest effort on my part to share the design for something I had already made for myself, I helped others. I may never meet them or benefit, but making the world a slightly better place is its own reward. That said, improvements built on your seed idea are reshared back with others and continue to be built upon. Best of all, these improvements, remixes, and mash-ups are all available for free to you as well. Who knows? Maybe I will need a water-based habitat sometime and directly benefit from the initial sharing of my fabric fastener. Honestly though, as captivating as Kevin Costner was in *Waterworld*, I hope not.

Amazingly, the sentiment that underlies modern open-source innovation is truly ancient. Even the Old Testament spoke to it in Proverbs 11:25: "A person who gives to others will get richer. Whomever helps others will himself be helped." To see how this can work even from modest sharing in our modern networked world, consider the cartoon in Figure 1.4. This cartoon explains how far sharing can spread using existing models (discussed in detail in this book). It tells the story of a simple act of sharing. You could literally do it with the phone in your pocket now. We can all make such an effort, and the cartoon shows how this simple act can expand far beyond the plans of the original sharer.

The story shown in Figure 1.4 is of a woman named Jen who used a series of life hacks (covered in Chapter 2) that simplified her life. With this saved time, she decided to go for a hike. On the hike, she noticed a single flower. It was a pretty flower. She pulled out her cellphone and snapped a picture of it. Then, to make sure she got a really good picture, she slowly walked around the flower and took a few dozen more

Figure 1.4 Cartoon of sharing (original art).

shots in rapid succession. The whole process took her a minute or two. Now normally she would have deleted all but the best picture and saved it for herself. But what if she chose to share those pictures instead?

Spending another few minutes to go to an online photo-sharing site (as discussed in Chapter 3) called Pexels (www.pexels.com), she uploaded and tagged her pictures "Flowers" and "Rose" and released them under a Creative Commons (CC0) license. This put them in the public domain so that anyone could use them for any purpose. Because Jen shared, others can now use the photographs in any way they like. Because she uploaded and tagged all of them, they are easy to find. They provide a range of options, so Jen's pictures were viewed and used by many people she may never meet.

The first person to see the pictures was a botanist from Melbourne working on improving a free online encyclopedia's (Wikipedia, wikipedia.org) articles on flowers. He used one of the photos there by posting it on WikiMedia Commons (commons.wikimedia.org) so that others would know what a specific variety of rose looks like. These images can also be used to train open-source image-analysis software and artificial intelligence programs to identify roses.

The Wikipedia article mentioned that all roses are edible plants. Another wiki editor, a Peace Corps student, decided to port the information to Appropedia (appropedia.org), a website devoted to sustainability and poverty reduction, to a page covering edible wild plants.

A chef, having read the Appropedia article, decided to use the image for her wild food book. Each recipe was accompanied by both a photo of the final mouth-watering dish, as well as pictures of the wild plants that made it. The chef wrote the book in Libre Office (www.libreoffice.org, a free office suite discussed in Chapter 6). In addition to selling the hard copy of the book, the chef posted the digital version for free under an open license on Freebooks (www.free-ebooks.net).

Because the chef shared her book for free, a couple decided to make an audio version. They recorded themselves cooking, making mistakes, and showing their friends trying the food. It was pretty funny. One of their friends took the audio, cleaned it up with Audacity (sourceforge.net/projects/audacity, a free audio mixer, covered in Chapter 5), and turned it into a full audio version and posted it on LibriVox (librivox.org, which provides free public-domain audio books).

Another writer, using Scribus (scribus.net), decided to use one of the other views of the rose in a picture book of flowers for children and posted it on Free Kids Books (freekidsbooks. org). A kindergarten teacher liked it so much that she printed it out for her students to enjoy in class.

Then an artist downloaded one of the photographs and used it to create a fantastic piece of digital art using the open-source digital painting program Krita (krita.org). He then shared it on Deviantart (deviantart.com, a community of online artists, covered in Chapter 4).

A band manager really liked the flower art and decided to use it on a concert poster for his band and posted it on the band's website. The poster was hung all around town. A music fan liked the poster so much that he used the image for a T-shirt and made a bunch for all his friends. The lead vocalist of the band wore the T-shirt at a concert. The band recorded its music from the concert and posted it on the Free Music Archive (freemusicarchive.org, covered in Chapter 5).

A disk jockey downloaded the mp3 of the songs, added them to a mix CD of free music, and then printed a copy of the flower picture to use as the cover of the CD case. Then things started to get out of hand. A cat owner decided to make a mini-me version of the flower T-shirt for her cat. Her boyfriend took a video (as discussed in Chapter 7) on his phone of the cat dancing to the band's music and posted it on YouTube, and it went viral. The cat's moves were so good that a movie graphics team used them as a model for creating a digital cartoon of a dancing cat using Blender (blender.org). The larger video was released by the Dutch Blender Foundation and used the music from the band in the scene.

A clothes designer saw the image on the band's website too and decided to make a dress out of it. The dress design was also shared for free, and then manufacturers (including home-based clothing printers, which we will meet in Chapter 9) began producing it at low cost. The manufacturers shipped the dresses all over the world and needed to set up pricing using free maps overlaid with geographic information system (GIS)–based shipping information (Chapter 8). Jen's mom bought her one for her birthday and put it in a wooden box that her father had made for her using open-source woodworking tools and techniques (Chapter 10).

Meanwhile, the other 29 photographs of the flower sat in a database. They didn't do anything for a while. Then a 3D designer downloaded them all—harvesting them like a bear eating blueberries in the same field by the rose. The designer used AliceVision and MeshRoom (alicevision.github.io) for free photogrammetry. This is a fancy way of combining photos to make a 3D object. First, he created a point cloud, which he then turned into a mesh and finally a stereolithography (STL) file. He 3D printed the flower using the STL file, and he gave it to his wife for their anniversary. She was really happy because he had made it specifically for her. It meant more to her than any store-bought trinket. She gave him a big kiss. He figured that he better pass the goodwill forward. He posted the 3D design (discussed in Chapter 12) of the flower on YouMagine (youmagine.com), a website from Ultimaker, a European open-

source 3D printer company. The 3D flower began to be downloaded a lot because it was so realistic. Many people used open-source electronics (Chapter 11) to drive their own homemade 3D printers (Chapter 12) to make physical copies. Computer-aided design (CAD) specialists from all over the world began to make various mash-ups of the 3D flower with things they were making. One of them made a curtain holder. Another made a pipette stand for scientists (Chapter 13) that is used by citizen scientists studying drinking water in their cities (Chapter 14). Still another made a rose-covered tablet computer holder. It took several tries, but the waste was reused with a recyclebot (Chapter 15) and converted back to feedstock for the 3D printers. Someone else made a vase with the rose all over it—and then a hydroponics system for his open-source house (Chapter 16). People from all over the world began downloading and sharing these designs, as well as printing them out as gifts. One of those people was Jen's husband, who printed out the vase and gave it to her on Valentine's day. Overall, the sharing of the pictures of the rose with the global Creative Commons community generated millions of dollars of value (Chapter 17).

The story ends with Jen at home in the flower art dress watching the cartoon cat video with the rose-inspired music on her open-source DLT1 tablet (published by Hackaday, hackaday.io/project/164845-dlt-one-a-damn-linux-tablet) and eating flower hors d'oeuvres she had read about in the cookbook. She has a bunch of roses next to her in a 3D-printed rose vase.

This is the future of how we all grow rich by sharing (Chapter 18). In each case in the cartoon, each person only made a small contribution. However, in aggregate, a single flower in a field had a major impact by inspiring people from all over the world to use their creativity to create value. Sharing enabled the value to scale the wealth laterally through many peers. This is an example of *peer production*. Everyone benefited from the sharing. Similar sharing moments are occurring all over the world as more people join the free and open-source community. As this happens, our collective wealth grows exponentially.

The benefits for individuals are clear, and it may be that people are joining simply out of self-interest. Those who join are following a well-established ethic first seen in the free and open-source software (FOSS) movement. The FOSS ethic was developed first by computer programmers sometimes called *hackers*.

One of the greatest gifts the FOSS movement has provided is a complete software ecosystem—all free. You can take advantage of this by getting far more life out of your current computer. At my university, computers are passed down from one user to another until they are so old they no longer work with Microsoft Windows. Then they are discarded as e-waste. We save as many as possible, and my lab is run primarily on these discarded computers because we resurrect them with free and open-source operating systems based on GNU/Linux. This saves us money—a lot of it—but also we get the performance we want because Linux is truly technically superior to Windows. You can do this too. There are hundreds of different kinds of Linux (called *distributions* or *distros*, distrowatch.com/), but for new users I recommend Linux Mint (www.linuxmint.com). Linux Mint is one of the most popular desktop Linux distributions and is used by millions of people because:

- It works out of the box, with full multimedia support.
- It is extremely easy to use.
- It is completely free of cost and open source.
- It is community driven. Users are encouraged to send feedback to the project

so that their ideas can be used to improve Linux Mint.

- It is safe and reliable because of a conservative approach to software updates.
- It requires very little maintenance (no regressions, no antivirus, no antispyware, etc.).
- It is based on Debian (www.debian.org), which is for the hardcore user, and Ubuntu (ubuntu.com), which is one of the other major distributions. This means that it provides support for a massive number of free programs—about 30,000 packages in all with supereasy-to-use software managers.

In the free Linux ecosystem, there is free software for just about anything you could want. Throughout the rest of this book, I will highlight free software that is particularly good for creating. If there is any other software you want that does not come with Linux, you can probably find an equivalent free software at OSALT (osalt.com). To get started with Linux Mint, simply follow the instructions on the following website to download it and make a bootable USB drive: https://linuxhint.com/install_linux_mint_19/. Then you can "try it before you buy it" by booting off of the USB drive and making sure you can use it. If all goes well, you can back up all your files and permanently install the software from the USB drive. Then load up all your personal files, and you have a state-of-the-art operating system. I am quite confident that you will see an immediate improvement in performance. It will be faster, and you will have at your fingertips any software you need simply by searching in the software manager for it and clicking the big green Install button—it all installs automatically for free. It just works, and we have the sharing community of thousands of FOSS developers to thank for it.

Through principles of free sharing and open access, open-source development treats users as developers. This is really clear among software developers who use their own code. This is because FOSS encourages contributions from everyone—as long as they are good. Good work is recognized and included in newer versions of the code. This has the effect of propagating superior software code. When we think about sharing beyond software, we simply expand the definition of *code* to mean the information needed to make anything from a song to a table rather than only computer software. This philosophy can be thought of as growing from the "white-knight hackers" who created the open-source movement with their *hacker ethic*, which embraces the following general principles (Levy, 1984, 2001):

1. Sharing
2. Openness
3. Decentralization
4. Free access, and
5. World improvement

The FOSS movement has produced a community of hackers and computer programmers whose shared goal is to work together to develop better computer software (DiBona, Ockman, and Stone, 1999). Remarkably, they have been outrageously successful! This philosophy of life works because it is supported by the gift culture of open source, in which recognition of an individual is determined by the amount of knowledge he or she gives away (Bergquist and Ljungberg, 2001). In such a system, *the more valuable the gift, the more respect you gain. Literally, the more you give, the richer you become.* Although such thinking may be reminiscent of the gift giving within the Hobbit communities in J. R. R. Tolkien's stories (2012), there are also modern analogues. Philanthropists know that the more they give away, the greater is their esteem. In addition, academics use this philosophy, which rewards

contributors through a process of peer review of scientific articles. The more good ideas academics share, the higher is the respect they command, the larger are their number of invitations to speak, and the more funds they can attract to continue their research. We have all had the experience of perhaps giving some advice to a stranger for free and enjoying the gratitude—and expecting similar help from others in return, no matter how small the request. When someone asks, "Where is the bathroom?", you will always tell them to help them out—and even if you are in another country, you can be sure "¿Dónde está el baño?" will get you some good-quality, free and timely information.

The reason this works so well is that all humans are actually hardwired to be nice. This may be hard to believe with the seemingly endless stream of terrible atrocities we read about in the news. There do seem to be a lot of them. If you think, however, about all the billions of people that every day manage not to commit any atrocities, there may be something to the "humans are nice" theory.

The Theory of Nice

We all prefer to hang around nice people. How many of your friends are not nice people—or at least nice to you? We do not need to read the scientific literature to know that nice people have more friends. It turns out that a huge number of studies go far beyond this to show that being nice is the way to go—not only for you, but for all of society.

For example, Harvard University's celebrated evolutionary biologist Martin Nowak (2011) explains that cooperation is the central key to the success of the 4-billion-year-old puzzle of life. Consider the following scenario: You are on the sidewalk in front of a store, and you notice a small child chasing a ball. The child is not looking and is about to enter the street, where he will almost certainly be hit by an oncoming car. Your natural reaction is probably to yell "Stop!" You might rush into the street to save the child. Good for you! But wait, why would you trouble yourself at all, let alone risk your own life, to run into the street to save a stranger's kid? You do it because your ancestors had a distinct evolutionary advantage to cooperate. Those selfish cave dwellers who did not help one another or cooperate got left alone by the clans to starve on the savanna. It is now clear that cooperation, not competition, is the defining human trait (Nowak, 2011). In fact, humans are not hardwired to lead lives that are "nasty, brutish, and short," although some started that way—those who made it all the way through are inborn to be good (Keltner, 2009).

Consider this: I am working on a building project at home while looking after my pre-preschool-age son. I drop a screwdriver on the floor, and my 2-year-old son grabs it and hands it back to me, without my asking, pointing, or otherwise telling him what to do. Now it is true that he is awesome, but it turns out that in this behavior, he is not special. You can run this same experiment in front of any young child and get the same result. This is not a learned behavior at all. Psychologist Michael Tomasello, who runs the Max Planck Institute for Evolutionary Anthropology in Germany, has shown through exhaustive observations of experiments on young children that children are helpful by default (Tomasello, 2009). Human children are naturally (and, I should mention, uniquely in the animal kingdom) cooperative. Other species just do not cooperate, even when they can. There are not a lot of cooperative bears. But even with animals that are closer to human mental abilities and DNA, the answer is the same. For example, when apes are put through similar experiments to those used with human children, they demonstrate the ability to work together

and share. However, they choose not to. This is why we do not currently live on the "Planet of the Apes." Ape selfishness has made them evolutionary losers that we go to visit at cages in the zoo. We cooperative humans, not the selfish apes, are the dominant species on the planet. As our children grow, their hardwired desire to help without expectation of reward sadly often becomes diluted and perverted by a selfish, materialistic culture.

This dilution is counterproductive to our success. Unfortunately, evolutionary theory of survival of the fittest has been misunderstood. Some people thought that we could transfer survival of the fittest between species (such as us and apes) and transfer it to social interactions among humans. Many people bought this theory, most notably those in some corporate cultures. Following this flawed theory, it is argued that concepts such as charity, fairness, forgiveness, and cooperation are evolutionary dead ends, which are soon to go extinct. Using this flawed theory, some people are encouraged to pursue their own self-interest at all costs. Anyone who has played a team sport like basketball, soccer, hockey, or football might immediately notice the flaw in this logic based on their practical experience. Amazingly, selfishness still holds incredible sway in our culture. Yet you do not even need to be an athlete to see something wrong with someone always hogging the metaphorical ball. You know from your relationships that a strategy of all-out selfishness does not work in the real world. Selfish people are normally considered "jerks" and ostracized in our society. The enlightened gambler thus bets on being nice. Believe it or not, being nice may also help you to find an attractive date. It turns out that both sexes consider that individuals who invest in altruistic acts are better candidates for long-term relationships (Moore et al., 2013). See any Hallmark movie for additional evidence.

The hardwired cooperative nature of humans, however, can also be a two-edged sword—particularly for small children. As our children become more aware of being members of a group, the group's mutual expectations can either encourage or discourage altruism and collaboration. In our very recent past, greed has even been elevated to a virtue. Remember that the 1980s have been called the "decade of greed." This has the unfortunate result of watering down our intrinsic cooperative nature, which has been built up through evolution and been fostered in nearly every culture until very recently (Gensler, 2004).

Yet we can overcome this bad programming and regain our cooperative nature. It comes out even in the worst times. As I write this book under a government-mandated "shelter in place" law during the COVID-19 pandemic, it is hard not to notice the good news among all the death and suffering. People from all over the world are donating their gifts to help others—many volunteering to risk their lives in the world's hot spots to care for the critically ill. There is also an explosion of free and open-source designs (Pearce, 2020a) meant to help others—including open-source ventilators (Pearce, 2020b), face shields (Prusa Research, 2020), and masks (Free Sewing, 2020), among others. Many different kinds of makers are fabricating personal protective equipment (PPE) for their local hospitals and helping each other in their communities. To see how widespread this type of sharing and caring is, consider just one group called *Helpful Engineering* (2020). This group has congregated to aid in the COVID-19 pandemic response by developing both open-source hardware and open-source software. The "helpful engineers" are working on a wide range of medical devices to create solutions that can be quickly reproduced and assembled locally worldwide. Although just starting out, the group has more than 2,500 registered volunteers, and its "Slack team" has grown extremely quickly to more than 9,000 by March of 2020. These makers are largely professional engineers, but other makers, such as homebound American volunteers, are sewing thousands upon thousands of face masks to help shield doctors, nurses, and everyone else from the coronavirus (Enrich, Abrams, and Kurutz, 2020). This is just one group of many. It is nothing short of amazing to see how large the distributed net around the world is and how the makers have rallied to meet the needs of the medical community.

Meet the Makers

Even outside this crisis, ways to apply our hardwired niceness can be found in the modern *maker movement*. Makers, for example, have a mutual expectation of collaboration. There are a lot of makers, and their numbers are swelling. For example, a recent estimate by *USA Today* puts the number of adult makers in the United States alone at 135 million, which is well over half (57 percent) of the American population aged 18 years and older. The maker movement itself is a relatively new, contemporary subculture. The maker movement now represents a technology-based extension of the more old-school do-it-yourself (DIY) culture.

Makers are often depicted on TV and in the movies. Legendary maker MacGyver from the 1980s is known for his talents for improvising technical solutions to insane problems—or perhaps, more modernly, Tony Stark, the Iron Man superhero from Marvel's *Avengers* movies, or Kaylee, the engineer from Sci-Fi's *Firefly* known for her own "MacGyverisms" to keep a spaceship afloat. People who make their own stuff—from dinner to clothes, from fixing their own cars to constructing their own staircases—this is what it means to be a maker.

These real-life makers enjoy engineering-oriented pursuits for fun, such as electronics, robotics, 3D printing, and the use of computer numerical control (CNC) tools, as well as more traditional activities, such as metalworking, woodworking, knitting, sewing, and traditional arts and crafts. They are all about sharing their love for the technologies—both new and old. The maker culture, however, stresses new and unique applications of technologies and encourages invention prototyping—and, of course, sharing and building on one another's work. There is a strong focus on using and learning practical skills, passing them on, and applying them creatively. If you are interested in meeting some makers, the best place is normally a local *hackerspace* or *makerspace*, which are community-operated physical places where people can meet and work on their projects (wiki.hackerspaces.org). Makerspaces are sprouting up all over the United States (see Figure 1.5). Makers are not just playing around, although both in fictional literature (see *Makers* by Cory Doctorow [craphound.com/makers] for an excellent fictional introduction to the

subculture) and in reading their blogs, it looks like fun. With free and open-source hardware (FOSH) makers starting from where the FOSS community has left off, there are burgeoning communities of hackers and makers working together to build everything from open-source cameras (which we discuss in Chapter 3), to videos (Chapter 7), to recycling (Chapter 15). Makers are even predicted to foster the next great industrial revolution (Taylor, 2012).

Concluding Thoughts

This chapter explored reasons why you would want to be more sharing. It is natural—you are hardwired to do it. Luckily, there are also extremely pragmatic and self-serving advantages to joining the open-source community. Cooperation provides both our species and our organizations, and you, with a distinct competitive advantage. It is important to remember that you do not need to be Elon Musk (the billionaire responsible for open-sourcing part of the Tesla electric car, which we will learn more about in Chapter 16) to make a significant

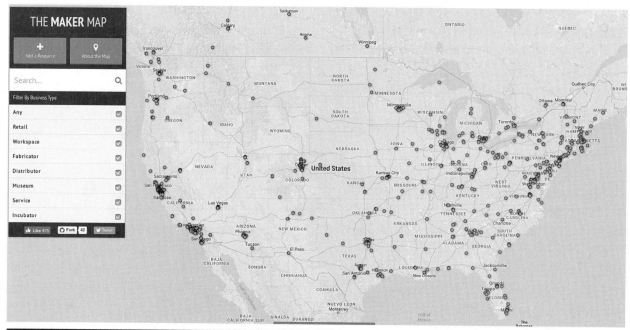

Figure 1.5   Screen shot of the MakerMap. For an updated version, see http://themakermap.com/

contribution. Every little bit of sharing has an impact and can snowball into something fantastic! To see some examples, next we will explore a large collection of free stuff you may want to enjoy, making projects of your own as compliments to those of seasoned sharers.

References

Bergquist M, Ljungberg J. 2001.The power of gifts: Organizing social relationships in open source communities. *Information Systems Journal* 11:305–320.

Enrich D, Abrams R, Kurutz S. 2020. A sewing army, making masks for America. *The New York Times*, March 25, 2020.

Free Sewing, 2020. Calling all makers: Here's a 1-page PDF facemask pattern. Now go make some and help beat this thing. https://freesewing.org/blog/facemask-frenzy/

Helpful Engineering. 2020. https://app.jogl.io/project/121#about

Keltner D. 2009. *Born to Be Good: The Science of a Meaningful Life*. W. W. Norton & Company, New York.

Levy S. 1984. *Hackers: Heroes of the Computer Revolution*. Doubleday, New York.

Levy S. 2001. *Hackers: Heroes of the Computer Revolution* (vol. 4). Penguin Books, New York.

Moore D, Wigby S, English S, et al. 2013. Selflessness is sexy: Reported helping behaviour increases desirability of men and women as long-term sexual partners. *BMC Evolutionary Biology* 13(1):182.

Pearce JM. 2020a. Distributed Manufacturing of Open Source Medical Hardware for Pandemics. *Journal of Manufacturing and Materials Processing* 4(2), 49. https://doi.org/10.3390/jmmp4020049

Pearce JM. 2020b. A review of open-source ventilators for COVID-19 and future pandemics. *F1000Research* 9:218. https://doi.org/10.12688/f1000research.22942.2

Prusa Research. 2020. 3D-printed face shields for medics and professionals. https://www.prusa3d.com/covid19/

Taylor C. 2012. *Wired*'s Chris Anderson: Today's "maker movement" is the new industrial revolution. *TechCrunch*. http://techcrunch.com/2012/10/09/wireds-chris-anderson-todays-maker-movement-is-the-new-industrial-revolution-tctv/

Tolkien JRR. 2012. *The Hobbit*. Houghton Mifflin Harcourt, New York.

Tomasello M. 2009. *Why We Cooperate: Based on the 2008 Tanner Lectures on Human Values at Stanford*. MIT Press, Cambridge, MA.

Weber S. 2000. The political economy of open-source software. Berkeley Roundtable on the International Economy Working Paper No. 140, Berkeley, CA.

CHAPTER 2

Making and Sharing Recipes, Life Hacks, and Household Tricks

Sharing Recipes

Nearly everyone can cook if they follow a good recipe. Now, with the advent of recipe-sharing websites such as Food.com (www.food.com) and Allrecipes.com (www.allrecipes.com), everyone has easy access to good recipes. These sites and others like them are active social networks of home cooks who connect and share recipes, photos of their culinary delights, cooking tips and tricks, and food trends. For example, if you decide to have spaghetti for dinner, Allrecipes provides not only tons of recipes for the spaghetti itself, but also lists recipes for side dishes to serve with it. Allrecipes already has far more recipes for specific dishes than any of the more conventional cookbooks (Teng, Lin, and Adamic, 2012). Gourmet meals from all over the world can be assembled easily in your own kitchen simply by following the directions. Thus, everyone with access to the internet can be a more passably good chef for an array of styles, eccentricities, and flavors than ever was possible before in history. Cooking at home from scratch saves money over processed meals or eating out, and in general, it is healthier for you as well. This saves you money in the long run on far more expensive medical bills. In addition, you can pull nutrition information from these sites, which can help you eat more healthily (if you want to). The beauty of open source and DIY is that *you have control*—you can add more or less salt based on your taste (or perhaps based on your salt sensitivity and high blood pressure).

As we learned in Chapter 1, it is not all about taking free stuff. If you create a new culinary masterpiece, you can add it to the collective taste bud wisdom by sharing it. Many of these sites have a rating system for the foods you can share. This allows users to get a general idea of the tastiness of a recipe to know if it is worth the risk. For example, Figure 2.1 shows a guacamole recipe with a five-star rating with 38 reviews and 100 makes. You can help others if you make the food by rating, providing feedback, or creating derivatives (perhaps with higher nutritional density). These kinds of projects are easy to get involved with, no matter what your skill level in the kitchen. For those just learning how to cook, adding a rating on something you made from someone else's recipe is a good start. You can also get your children involved in testing recipes from the web and trying to make their own culinary masterpieces. This is a good life skill for them that will save them money and benefit their health far into the future. If you know your way around a kitchen, consider publishing some of your best recipes to get feedback, hints for improvements, or drink matches from foodies all over the world.

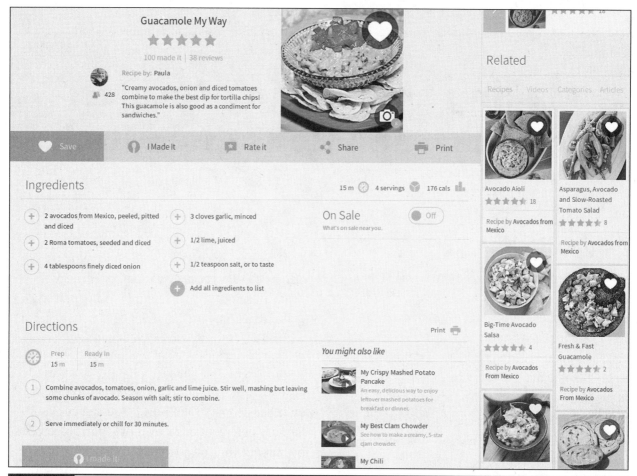

Figure 2.1 A guacamole recipe made with creamy avocados, onions, and diced tomatoes, combined to make the tortilla chip dip developed by Paula (Olive Branch, MS). http://allrecipes.com/recipe/237854/guacamole-my-way/

Sharing Life Hacks and Home-Improvement Tricks

As Gina Trapani (2008) explains, every day you have dozens of opportunities to go about your day faster, smarter, and more efficiently. You can do this by taking the right shortcuts. These shortcuts are often called *life hacks*. Life hacks describe advice, resources, tips, or tricks that will help you get things done more efficiently and effectively. Life hacks range from putting a rubber band around a paint can on which to wipe your brush and avoid drips, to applying eyeliner with a spoon to get a perfect curve. The internet community has already developed a mature culture of sharing life hacks (on sites such as Lifehack [www.lifehack.org], which provides tips to help improve your productivity in all aspects of your life). Life hacks include faster, easier ways to fold laundry (see ninja T-shirt folding) to putting a lazy Susan in the refrigerator to provide easy access to all your condiments. Small-scale genius ideas have been shared by people like you to make all our lives easier.

You probably have developed some life hacks yourself over the years. If you have a life hack you want to share, you can easily contribute it at Lifehack, although there are many good websites for discussing life hacks. For example, Reddit (www.reddit.com) is a discussion-based website that allows registered community members to submit content, such as text posts, images, or direct links to interesting content on the internet. Users vote submissions up or down to organize the posts and determine their position, where the submissions with the most positive votes appear on the front page or at the top of a category or interest, also called a *subreddit*. Reddit has a very active life hacks subreddit (www.reddit.com/r/lifehacks). Thus, for example, several Reddit users provided examples of three ways to fill a bucket from a sink if the bucket does not fit in the sink, as shown in my collage in Figure 2.2: Use a few Styrofoam cups, a water bottle, or a dustpan.

Another great source of life hacks is Instructables (www.instructables.com). Instructables is a website developed by Autodesk that holds user-created and -uploaded DIY projects. Not only can you post your own projects to share with the world, but you can also comment on other users' projects and rate them for quality. This gives you the opportunity to get valuable feedback on your own projects for free. In general, these project page descriptions are well organized and easy to follow, as shown in Figure 2.3 for a project on making an open-source solar-powered USB charger.

The range of life hacks is vast. For example, if you have made inexpensive cleaners from common household items (e.g., vinegar and water), you can help these efforts by performing controlled experiments on cleaning disasters around your own house and posting feedback on other people's DIY posts. You can cut your laundry detergent costs by more than 50 percent by making your own from shared recipes of borax, washing soda, and bar soap, along with nonactive ingredients, to make your clothes smell good. When everyone helps a little bit like this, our collective wisdom moves forward. With each bit of knowledge shared, you enable others to avoid the cost of buying price-escalated cleaning supplies, for example, which often contain chemicals with unfortunate health and environmental impacts (Venhuis and Mehrvar, 2004; Bello et al., 2009). Again, when you make your own cleaning solutions, if you don't want toxic chemicals in your house, simply don't add them.

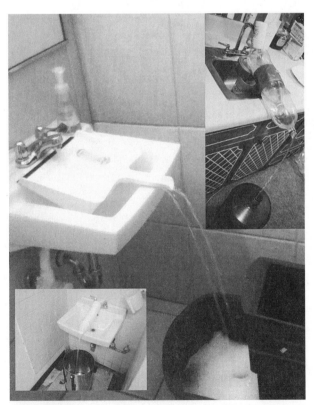

Figure 2.2   A collage of various life hacks discussed on Reddit to fill a bucket with water if it is too large to fit in a sink, including using a dustpan, taped together 2-liter bottles, or Styrofoam cups with holes poked in them.

Figure 2.3   Screen capture of Instructables guide to building an open-source solar pad kit solar USB charger. (CC BY-SA) http://www.instructables.com/id/DIY-Solarpad-Kit-Powerful-USB-Solar-Charger/

Open-Source Ad Blockers

My personal favorite open-source life hack has to be the ad blocker. If you do not use one of these already, making the switch to an ad-free internet compliments of open-source software will change your life! I have been surfing the web for years without ever seeing any ads, so it is always shocking to me when I get on someone else's computer and am virtually attacked by a screen full of garbage. These ads take extra time to load when you go to a webpage, so I was curious to know exactly how much time it wastes. In a study I just completed (Pearce, 2020), the results were astounding. The average

internet user would save over 100 hours a year by using uBlock Origin (github.com/gorhill/uBlock), a free and open-source ad blocker. This is nearly 2 hours a week! uBlock Origin was the most effective ad blocker tested, but all ad blockers save time, energy, and money. The results of my little study show that page load time dropped 11 percent with AdBlock+ (adblockplus.org), 22 percent with Privacy Badger (privacybadger.org), and a whopping 28 percent with uBlock Origin. These savings are not significant on a single page, but internet users spend more than half their time online rapidly clicking through websites, spending fewer than 15 seconds on a given page. With all these clicks, the additional time to load ads really starts to add up.

Internet users spend more than 6.5 hours per day online. Americans, for example, have more than doubled the time they spend online since 2000, to almost 24 hours a week. Open-source ad blockers have the potential to reduce the time, and thus electricity spent, by eliminating ads during internet browsing and video streaming.

In my study, three open-source ad blockers were tested against a no-ad-blocker control. To get a feel for the impacts of ads on your web-browsing experience, look at Figure 2.4. As you can see, a typical webpage is covered with big ads (Figure 2.4a), and AdBlock+ shrinks them (Figure 2.4b), Privacy Badger shrinks and replaces them with white space (Figure 2.4c), and uBlock Origin completely annihilates them, making more of the content of a page visible (Figure 2.4d).

Page load times were recorded for browsing a representative selection of the globally most-accessed websites, including web searching (such as Google, Yahoo, and Bing), information (such as weather.com), and news sites (such as CNN, Fox, and *The New York Times*). In addition, the study analyzed the time spent watching ads on videos for both trending and nontrending content. This part of the study was more challenging because of the lack of data on what ratio of YouTube watching time is spent on trending and nontrending content. The time wasted viewing ads per video ranged from 0.06 percent up to a staggering 21 percent. Thus, total hours lost to loading ads was only recorded for browsing.

Overall, the results showed that the energy wasted loading ads is not trivial. Because a lot of the electricity used for running computers continues to come from coal, which causes

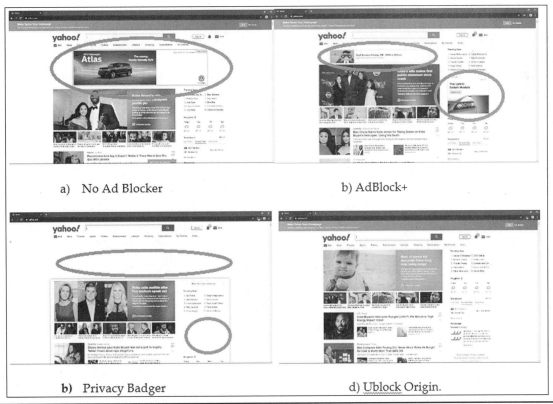

a) No Ad Blocker   b) AdBlock+

b) Privacy Badger   d) Ublock Origin.

Figure 2.4 The impacts of open-source ad blockers on internet web browsing: (a) no ad blocker versus open-source ad blockers; (b) AdBlock+; (c) Privacy Badger; (d) uBlock Origin.

air pollution and premature death, the study analyzed the potential for ad blockers to save American lives. The results were shocking: The energy conserved if everyone in the United States used the open-source ad blocker would save over 36 Americans lives per year.

Electricity costs money, so cutting ads could also save consumers money. In the United States, if all internet users enabled Privacy Badger on their computers, Americans would save more than $91 million annually. Globally, the results of the investigation were even more striking. uBlock Origin could save global consumers more than $1.8 billion per year.

Although free and open-source ad blockers save energy and are good for the environment, you are probably going to use them primarily to block annoying ads and save yourself time. You can then use that time to work on your own open-source projects!

References

Bello A, Quinn MM, Perry MJ, Milton DK. 2009. Characterization of occupational exposures to cleaning products used for common cleaning tasks: A pilot study of hospital cleaners. *Environmental Health* 8(1):11. https://doi.org/10.1186/1476-069x -8-11

Pearce JM. 2020. Energy conservation with open-source ad blockers. *Technologies* 8(2):18. https://doi.org/10.3390/technologies8020018

Teng CY, Lin YR, Adamic LA. 2012. Recipe recommendation using ingredient networks. In *Proceedings of the 4th Annual ACM Web Science Conference* (pp. 298–307). ACM, New York.

Trapani G. 2008. *Upgrade Your Life: The Lifehacker Guide to Working Smarter, Faster, Better*. John Wiley & Sons, Hoboken, NJ.

Venhuis SH, Mehrvar M. 2004. Health effects, environmental impacts, and photochemical degradation of selected surfactants in water. *International Journal of Photoenergy* 6(3): 115–125.

CHAPTER 3

Making and Sharing Digital Photographs and Open-Source Cameras

Taking and Sharing Digital Photographs

We all take pictures, and many of us are already used to sharing them with our friends and family on social networks, as in Figure 3.1. However, to enable any photograph you take to be free and open source, whether with a professional camera or with your smartphone, you need to upload it to the Creative Commons. In other words, it must be uploaded to a website that allows you to share it with a license that allows others to freely use it. This is far less challenging than it sounds. There are many websites that make this easy, and here we will look at just three.

Figure 3.1 A man and a woman taking a selfie. (CC0) https://www.pexels.com/photo/ photo-of-man-and-woman-taking -selfie-1371176/

Pixabay (www.pixabay.com) allows you to upload/download royalty-free stock photos, vector graphics, illustrations, and videos and to share your own photographs in the public domain (Creative Commons zero [CC0]) with people all over the world. You can choose from a large selection of public-domain images; download them in small, medium, large, or original (supersized) formats; and use them any way that you would like. Pixabay allows comments, tags, and categorization. As of this writing, it has more than 750,000 images from which to choose.

Pexels (www.pexels.com) is another website to share your digital images. It similarly follows the CC0 license, which means that the pictures are free for personal and even for commercial use. You can modify, copy, and distribute the photos without asking for permission or setting a link to the source. In addition to the size choices from Pixabay, Pexels also lets you input a custom size so that you can download exactly what you want. There are dozens of other websites that offer similar images, such as Public Domain Pictures (www.publicdomainpictures.net). In addition to presenting a standard sharing platform, the site also has compiled a selection of photography tutorial videos that may be of great help to a new photographer. Several of the illustrations in this book (e.g., Figure 3.1) and in many other books have come from talented photographers who have provided images on Pixabay and Pexels.

Digital photographs have real value. This value becomes apparent to everyone who has to pay for it. For example, Getty Images, a commercial stock photography website, charges $175 to $575 for a single image depending on the size (although special images go for far more). People all over the world need good photos that are appropriately tagged for decorating websites, writing reports and books, and engaging in other creative acts. As Chapter 4 will show, there are even several free and open-source programs that enable you to take photographs and then turn them into other forms of digital art. The more material these creative people have to work with, the better are their products. The more others share, the better your projects will be too! Much of the work we all create will be fed back into the commons and result in more value for you.

There are several incentives that such photo-sharing websites use to encourage uploading. These include micropayments, tips, monthly contests, peer recognition, download stats as a vehicle for personal pride, and the ability to increase social media following, which comes with its own list of benefits (e.g., you could be a paid "influencer"). The simplest websites at least enable you to track the use of your images. You can also do this manually using a Google image search to see how often your pictures are used on the web. Go to Google Image search, and then drag and drop your picture into it. The results will show you the other places that your picture shows up on the web.

You are probably already well aware of many social media sites that enable you to share photographs. Most users of social media websites have never looked at the fine print on exactly what the licenses are for sharing their content. If you post a picture on Twitter, who owns the copyright? Did you give it away? Can others use it? What about Facebook? LinkedIn? Or the other social websites? To get a handle on the default settings for photo-sharing sites, please see Table 3.1. Sadly, as you can see in this table, most of the social media sites have somewhat obtuse licenses. It would be far better if they specifically classed them in the public domain or made a Creative Commons license the default. A few of the social media sites allow you to select this option, but the vast majority of users are subject to the default and somewhat unclear license.

Table 3.1   Default Licenses for Social Media

Website	Number of Users, 2019	Default License	Notes (from license page of each website)
Facebook (www.facebook.com)	1.500,000,000	Content specific	"Content you share on Facebook is your intellectual property, though Facebook may share it as needed (consistent with your privacy settings)."
YouTube (www.youtube.com)	1,499,000,000	Standard YouTube license, but a CC license is available	"When a user is uploading a video, he has license options that he can choose from. The first option is the 'standard YouTube License,' which means that you grant the broadcasting rights to YouTube. This essentially means that your video can only be accessed from YouTube for watching purposes and cannot be reproduced or distributed in any other form without your consent."
Twitter (www.twitter.com)	400,000,000	Public domain (sort of)	User retains all rights to content published to Twitter, but Twitter gains a license to freely share and distribute it and to give that same sublicense to other users.

Website	Number of Users, 2019	Default License	Notes (from license page of each website)
Instagram (www.instagram.com)	275,000,000	Content specific	Same license as Facebook because it is owned by Facebook
LinkedIn (www.linkedin.com)	250,000,000	Typical social media license but a CC license is available	LinkedIn has nonexclusive rights to freely share and sublicense your content until your content's/account's deletion.
Reddit (www.reddit.com)	125,000,000	Public domain (sort of)	Nearly the same license as Twitter, though seemingly more freedom is given to Reddit. We will call this the "vague social media public domain license" because it does not specifically say public domain but allows republishing and alterations.
VK (www.vk.com)	120,000,000	Full rights to content owner	"After registration, the User obtains the right to create, use and determine independently for personal purposes the content of his/her own personal page and conditions for other Users' access to its content."
Tumblr (www.tumblr.com)	110,000,000	Public domain (sort of)	Vague social media public domain license
Pinterest (www.pinterest.com)	105,000,000	Public domain (sort of)	Vague social media public domain license
Google Plus (www.plus.google.com)	100,000,000	Public domain (sort of)	Vague social media public domain license
Flickr (www.flickr.com)	80,000,000	View/share-only license	Content published to Flickr is public, though no modification of the content is allowed.
MeetUp (www.meetup.com)	42,000,000	Public domain (sort of)	Content is still yours if you share it on MeetUp, but MeetUp has the usual social media license, and your content remains on the platform after deletion/closure of your account.
Ask.fm (www.ask.fm)	40,000,000	View/share-only license	Content published to Ask.fm is public, though no modification of the content is allowed.
LiveJournal (www.livejournal.com)	37,000,000	Public domain (sort of)	"In respect of any Content which constitutes intellectual property, User provides to the Administration a non-exclusive (simple) license to use his/her Content in order to provide the Service by reproducing his/her Content as well as by making it public for the entire period the Content is posted on the Service. If User participates in any rankings or if User's Content is used in any editorial projects of the Service, User provides to the Administration an additional authorisation to modify, shorten and amend his/her Content, to add images, a preamble, comments or any clarifications to his/her Content while using it, and in certain cases based on the Service functions, an authorisation to use User's Content anonymously."
myspace (www.myspace.com)	10,000,000	Public domain (sort of)	Typical social media license, though efforts are taken after content/account deletion to remove your content from the site.

Free Software for Photo Lovers

There is also a lot of open-source software that can help you take and finish your pictures. For example, DigiCamControl (digicamcontrol .com) is open-source software that helps you control your camera settings remotely from your computer. You can hold the camera, shoot, and have the resulting images displayed on the computer monitor.

If you need to improve your photographs before you post them, the Gnu Image Manipulation Program (GIMP; www.gimp.org) is free, open source, and relatively easy to use. GIMP can do professional cropping, color correction, brightness control, contrast correction, and many other standard image-processing routines. GIMP also comes with a long list of artistic filters that can be extended to provide for full-featured photo touching. For example, Figure 3.2 shows the before and after colorization with GIMP of the Great Depression photo "Migrant Mother" by Dorothea Lange. GIMP even has powers to make your vacation pictures much better. If you have been to any tourist destination recently—really anywhere in the world—you have probably noticed that there are too many tourists! This phenomenon now even has a name—*overtourism*. Locals in many tourist areas such as Europe and other popular destinations are fighting back and trying to cut down on tourists. This doesn't help you when there are still dozens of people standing in the way when you want to get a picture of a famous statue, building, landmark, artwork, or natural wonder. GIMP to the rescue! You can use GIMP to make your photographs tourist-free with an open-source plugin called *G'MIC*. First, put up a tripod and take an odd number of pictures every 10 seconds or so for a few minutes as the hordes of tourists mill about. Load up each of your aligned images as a layer in GIMP. Run

G'MIC. Then go to Layers, Blend [median], and make sure that the input layers are all visible. Then, finally, output to a new layer. This will take a bit of time, but you end up with a tourist-free shot of your favorite destination or sight—just like in the brochures!

(a)

(b)

Figure 3.2   Before (a) and after (b) colorization with GIMP of Great Depression photo "Migrant Mother" by Dorothea Lange. Colorized by John Boero. (Public domain) https://commons.wikimedia.org/wiki/File:MigrantMotherColorized.jpg

Open-Source Cameras

Open source doesn't have to stop with the images or software to sort and edit them. Even the tools you use to take your photos can be open source and fully hackable. Elphel, Inc. (www.elphel.com) designs and manufactures high-performance cameras based on free software and hardware designs. The GNU General Public License (www.gnu.org/licenses/gpl.html) and the CERN Open Hardware License (www.ohwr.org/projects/cernohl) cover all the Elphel software and hardware designs. With open hardware camera companies such as Elphel, you have the option to buy the product that works out of the box like a normal purchase. Even better, you also have the possibility, and information needed, to modify any parts inside them. Elphel cameras have been used from everything from helping to capture images for Google Street View to taking pictures under the sea ice for surveying and exploring in Antarctica. The company has crazy cameras in its lineup. Consider the Elphel Eyesis4Pi-26-393, which is a full-sphere multicamera system for stereophotogrammetric applications, as shown in Figure 3.3. The free and open-source software that comes packed with it compensates for optical aberrations and preserves full resolution of the sensors over the field of view. You can use this for precise pixel mapping to automatically stitch images into breathtaking panoramas, which you may be familiar with. But it also allows for easy photogrammetry, which allows you to stitch your two-dimensional (2D) pictures to fabricate accurate three-dimensional (3D) reconstructions. You can walk into a building with this camera and walk out with a full 3D model of every room you enter. These 3D panoramas can be useful for everything from creating real settings for video games you are creating to giving your mom a virtual tour of your apartment when you are overseas.

Open-Source Software for Photogrammetry and Machine Vision

You do not necessarily need a fancy camera to enjoy the benefits of photogrammetry.

Meshroom (alicevision.org/#meshroom) is a free, open-source 3D reconstruction software based on the AliceVision framework. In the not-so-distant past, photogrammetry was extremely complex and hard to do. What Meshroom does is make it easy to do 3D reconstructions, photo modeling, and camera tracking just by taking a bunch of pictures with an ordinary digital camera or smartphone camera. The state-of-the-art computer vision algorithms that AliceVision

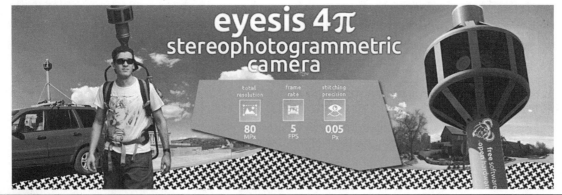

Figure 3.3 Elphel Eyesis4Pi-26-393, a full-sphere multicamera system for stereophotogrammetric applications. (CC BY) https://www.elphel.com/wiki/Eyesis4Pi_393

uses are all open source, and you can dive into the code as deep as you like—but for most users you just need to follow some simple steps to create fantastic 3D models from just a few pictures. The models can be used to fabricate things (e.g., 3D printable models, as in Chapter 12) or to make ultrarealistic backgrounds for your open-source games or computer-generated imagery (CGI) movies. With your object being still (not moving), you take at least 30 (more is better) pictures from different angles and heights. You load all the images into Meshroom and hit Start so that it begins to match up the camera angles and build a 3D model of what you are aiming at (e.g., the landscape, person, and work of art examples shown in Figure 3.4).

Maybe you want to take pictures without being there, and you have had enough technical experience to be able to jump into a little computer code. There are several project options available in the open-source community for you. The OpenMV (openmv.io) project is creating low-cost, extensible, Python-powered, and machine vision modules for makers and hobbyists. OpenMV Cam is like an Arduino (www.arduino.cc) with a camera on board that you program in Python. It is easy to run machine visions algorithms on what the OpenMV Cam sees, so you can track colors, detect faces, and more in seconds and then control input-output pins in the real world.

(a)

Figure 3.4 Screenshots of Meshroom examples: (a) Davia Rocks in Corsica. (AliceVision)(CC BY) https://forum .sketchfab.com/t/open-source-photogrammetry-with-meshroom/22965/5

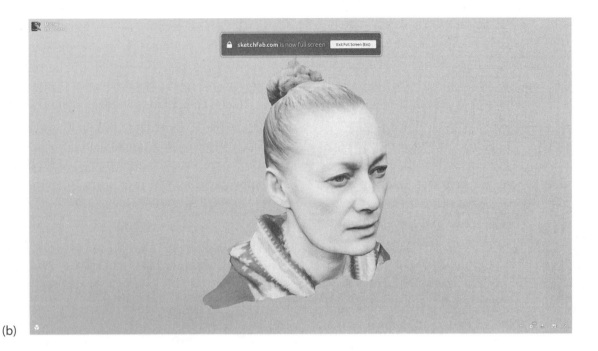

(b)

(c)

Figure 3.4 Screenshots of Meshroom examples: (b) Mother (Meshroom with final touchups in Blender by Vincurek.f); (c) Saint Guilhem Cloiser Foundation (80 images processed with default settings by Thomas Flynn). (CC BY) https://forum.sketchfab.com/t/open-source-photogrammetry-with -meshroom/22965/5

If you want to learn more about computer vision (CV), the Open Source Computer Vision Library (OpenCV, opencv.org) is the preeminent open-source CV and machine-learning software library. The OpenCV community is particularly diverse and rich, with more than 47,000 people in its user community and tens of millions of downloads all over the world. There are probably others out there that have tackled a similar CV project to the one you might be interested in with OpenCV. Generally, they are glad to pass on tips because they were helped by others in the community. OpenCV has a very permissive free software Berkeley Software Distribution (BSD) license, which imposes minimal restrictions on use and redistribution. This makes it really easy for businesses to use and modify the OpenCV code—so it is used all over the place.

The thousands of optimized algorithms that make up the OpenCV library help you to easily do the following:

1. Detect and recognize human faces
2. Identify objects (e.g., cars)
3. Classify human actions in videos
4. Track camera movements
5. Track moving objects (e.g., pucks during a hockey game)
6. Extract 3D models of objects or produce 3D point clouds from stereo cameras (which is handy for making the 3D printable objects we will talk about in Chapter 12)
7. Stitch images together to produce a high-resolution image of an entire scene for amazing vacation panoramas
8. Find similar images from an image database (such as the Google image search)
9. Remove those pesky red eyes from family photos taken using a flash
10. Follow eye movements (e.g., to help learn what people are looking at to make better education aids or to prevent car accidents from drowsy drivers)
11. Recognize scenery
12. Establish markers to overlay the photo with other software such as augmented reality

You can absolutely play with and use OpenCV like your own personal robot eye toy, but major companies—think Google (e.g., stitching streetview images together), Yahoo, Intel, IBM, Sony, Toyota, and many others—all use OpenCV. Many companies and individuals have contributed to making the algorithm libraries even more robust and available in C++, Python, Java, and MATLAB interfaces.

These open-source CV projects are maturing rapidly and becoming easier to use. Soon there will be apps that anyone can use with even the most modest computer experience. Today, you still need to be comfortable with a bit of programming—but luckily, the open-source community has provided oodles of tutorials that can train you in these tasks for your projects.

CHAPTER 4

Making and Sharing Art

One can make art for oneself, but the vast majority of it is made with the express purpose of being shared with others. Because of this, there is a thriving online art community—as if you lived in Bohemia during its height of art inspiration. Artists can meet online and share their work; give each other comments, constructive criticism and critiques, and motivation to improve; teach each other; and be inspired by each other's work. Some artists even share their art in steps so that others can learn from it. This is particularly helpful when you are just learning to draw or paint. Perhaps most important, being surrounded (albeit virtually) by other like-minded artists can be a lot of fun.

The largest online community is Deviantart (www.deviantart.com), which is set up to nurture artists, from newbies to professionals. Deviantart opens up with new art and undiscovered artists, such as that shown in the September 2016 screen shot in Figure 4.1. Each piece of artwork you share automatically comes with a message board so that other artists can comment on your work and you can enable the notification system. It also lets you collect and organize inspiration from others. Every new user gets his or her own homepage, which can be customized nicely with various widgets to show off your work in whatever way you would like to arrange it. You can follow other users or follow their deviations (images), journal posts, and forum posts to allow you to tune your interest level in others. Finally, the site enables you to develop or join existing groups of other artists with similar interests (e.g., techniques or genres) and share and collect artwork. This makes it easy to collect a bunch of cool images to use as the background for your computer, but it is probably best for finding inspiration for your own work as you grow as an artist. Licensing is treated fairly loosely on this website—one should assume that most of it is copyrighted, although Creative Commons licenses can be enabled. The Deviantart community is large and vibrant.

Another good art community that offers honest critiques of your art, if you ask for them, is Concept Art (www.conceptart.org). Although perhaps not as sensitive to your feelings (be tough!), the forum website is information dense. When you join, you make a personal sketchbook thread with a showcase gallery of your work. If you really want to learn and improve your art skills, you can also take fine-art classes called "Level Up." In addition, the site hosts community activities and challenges, as well as ongoing art discussions.

There are also open-source sites devoted to specific kinds of art, such as Open Game Art (opengameart.org), which helps if you need some art to make your video game cooler—or if you want to help others with their games. One user commented that she likes that there are lots

Figure 4.1   Screen shot of some art posted for comment by the Deviantart community.

of easily accessible resources for when she gets stuck, no matter how complex her art project.

If your art interest focuses on computer graphics, CGSociety (www.cgsociety.org) may be a great place for you, whether you are into 2D or 3D art (if you just need some free clip art, then FreeSVG [freesvg.org] has tens of thousands of images for you to look through). CGSociety has a large repository of articles on the digital entertainment industry, their artists, and what is going on in the latest movies and short films. You can post your work in the forums and get good feedback on technical questions and software such as Blender (www.blender.org). Blender is a free and open-source 3D creation suite. Frankly, Blender is crazy! It has everything you need to make a full-scale computer graphics (CG) 3D movie, including modeling, rigging, animation, simulation, rendering, compositing,

and motion tracking, even video editing and game creation. For example, consider *Sintel* (Figure 4.2), which is an independently produced short film initiated by the Blender Foundation, that was initially funded by thousands of internet donations. You can watch the 15-minute film for free (durian.blender.org), and download any of the raw files you need to make a derivative of your own. You can use the characters, backgrounds, scenes, and all the tools that made it because they are also open source.

Members of the open-source sharing community build off of each other. Having access to free software can really expand your horizons. For example, you can doodle or have some fun with your kids using TuxPaint (www.tuxpaint.org), which is an award-winning free drawing program for children ages 3 to 12 (preschool to sixth grade). TuxPaint combines

Figure 4.2 Blender-created image by Lee Salvemini, Pablo Vazquez, David Revoy, Durian Team, and the Blender Foundation. (CC BY) http://durian.blender.org/wp-content/uploads/2010/09/sintel_cover.jpg from http://durian.blender.org/news/sintel-on-3d-worlds-october-issue/

an easy-to-use interface, fun sound effects, and an encouraging cartoon penguin mascot (the Linux penguin Tux), who guides you as you use the program.

However, if you are going to do digital art, why not go all the way to the top and use a tool for professionals? You can do this with free and open-source software. For those of you interested in digital art, Krita (krita.org) is a

fantastic art program that is totally free. The Krita community aims to make Krita the best painting application for cartoonists, illustrators, and concept artists, and if you spend even a few minutes playing with it, you can see it's come a long way.

As Jack the Vulture3, a 22-year-old artist, explains, he likes digital art because it offers "So much more freedom. To experiment, to make mistakes, to change things around, to try whatever you can think in your mind without having to make the journey to an art supply store. Especially when you don't have the money to buy all those paints and canvases." This makes art accessible to anyone with a tablet or a computer who's willing to work to create. As Jack points out, "I won't run out of digital canvases" (krita.org/en/item/interview-with-jack-the-vulture/).

Like most of the best open-source packages, Krita is very customizable. There are scores of brush engines to play with, and a constant deluge of new features is being worked on all the time. Plus, you can even make your own brushes, and the developers are responsive and may include your recommendations in the next update. Krita user David Revoy shows how an open-source free culture might work. He is a fantastic illustrator (as shown by his tutorial that made the image in Figure 4.3).

Revoy also started the Pepper & Carrot project (example in Figure 4.4), a comedy/humor webcomic suited for everyone of every age. Somewhat refreshingly for those with children, there is no mature content and no violence, so you can safely read it to your kids without checking it first. Pepper & Carrot is a free and open-source comic about a little witch and her cat. It is funded only by its patrons; each patron sends a little money for each new episode published and gets a credit at the end of the new episode. Thanks to this system, Pepper & Carrot can stay independent and never has to resort to

Figure 4.3   *Aqua Dream* by David Revoy. Speed painting using Krita recorded for Youtube (youtube/dsDgtlWXLXE). (CC BY-SA) www.davidrevoy.com

advertising or any marketing pollution. Without any intermediary between artist and audience, users pay less, and the artist benefits more. All the content produced for Pepper & Carrot (www.peppercarrot.com) is free and available to anyone, and it was made with free/libre open-source software on GNU/Linux. Following the true open-source spirit, commercial usage, translations, fan arts, prints, movies, video games, sharing, and reposts are all encouraged!

For example, a large group of awesome artists made Pepper & Carrot animation using Krita, Blender, and several more free and open-source software (FOSS) tools, including (1) Papagayo, a lip-syncing program designed to line up mouth shapes (phonemes) with recorded speaking, (2) RenderChan, a tool for managing rendering of animation projects, and (3) Synfig Studio (www.synfig.org) for 2D animation. These packages are part of the Morevna Project (morevnaproject.org), which is creating open-

source animation shorts and developing FOSS tools for animation for everyone to enjoy and use. It also hosts the Pepper & Carrot animation. Another fun FOSS approach to animation is MonkeyJam (monkeyjam.org), which is a digital pencil test and stop-motion animation program. You capture images from a webcam or your phone and then assemble them as separate frames of an animation.

If you are looking for open-source tools that will help you do professional work, there are two more programs that may interest you. First, Inkscape (inkscape.org) is a professional vector graphics editor. Inkscape can help you create a wide variety of graphics, including illustrations, icons, logos, diagrams, maps, and web graphics. In addition, developers have freely provided a wide range of extensions that can help you develop more complex art, as well as interact with digital tools. For example, Inkscape helps you do things like a quick design for laser cutting

(which we will discuss in more detail in Chapter 12)—everything from joints to jigsaw puzzles to an extension for doing embroidery.

The other major powerhouse for professional art is GIMP (www.gimp.org), which is the GNU Image Manipulation Program we discussed in Chapter 3 for touching up your photographs. With a raft of customization options and third-party plugins, you can take images and transform them into real works of art, as well as make icons and interface art for other programs. For example, consider Figure 4.5, which was made in GIMP and used MathMap (complang.tuwien.ac.at/schani/mathmap) to turn a simple clock picture into a work of art. Last but not least, if you need to do a lot of image manipulation, GIMP is scriptable in many languages, including C++ and Python.

Figure 4.5   GIMP used to make art: *Back in Time for February…just* by Nikk. (CC BY-2.0) https://www.flickr.com/photos/nikkvalentine/39289966044

CHAPTER 5

Making and Sharing Music, Software, and Instruments

Great Places to Share Your Music

There are many places where artists can share their music in an open-source way, such as the Free Music Archive (freemusicarchive.org). This site is exactly what it is says it is—the best place to find free music and royalty-free music. The Tribe of Noise (artist.tribeofnoise.com) now owns it. The Tribe of Noise is a global Creative Commons music community made up of tens of thousands of independent musicians. Members aim to support individual music makers, while at the same time managing the combined Tribe talent to deliver music services to businesses. The Tribe of Noise blog lists opportunities for musicians to get greater exposure and get paid to create open-source music that is streamed on the Tribe's platform for businesses.

You can also share your music with friends using Funkwhale (funkwhale.audio), which is a community-driven project where you can listen to and share music and audio within a decentralized open network. Funkwhale gets its name because, just like whales, Funkwhale users gather in "pods." These pods are simply websites running free Funkwhale server software. You join the network by registering an account on a pod. This is your home, but you can interact with other users in any pod. Funkwhale is designed for both artists and music lovers. Artists, musicians, and podcasters can publish

work freely, reach a wider audience using the federation, and embed and share content on other platforms, websites, and podcasting applications. People who just want to listen to music can freely join a pod, upload their own music library on their pods, share their libraries with friends and family, and access their content from multiple devices. Perhaps best of all, Funkwhale can help you discover new music and artists because it operates as an open-source music streaming and social media hybrid.

Whether you are jamming to your own tunes or someone else's free music, you can even do so on open-source adjustable-headband high-quality over-ear headphones (Figure 5.1) that are easily made on a desktop RepRap-class open-source 3D printer (more details in Chapter 12).

Figure 5.1 EQ-1 headphones, high-quality adjustable-headband over-ear open-source headphones. (CC BY-SA) https://www.thingiverse.com/thing:674179

Open-Source Music Software

There is a lot of free and open-source software that can help you create music. Both Audacity (www.audacityteam.org) and Let's Make Music (lmms.io) are easy-to-use cross-platform multitrack audio editors. They are rich applications (as seen in the screen shot of Audacity in Figure 5.2). For hardcore music hackers, Qtractor (qtractor.sourceforge.io) is an audio/Musical Instrument Digital Interface (MIDI) multitrack sequencer application. It is used in Linux along with the Jack Audio Connection Kit (JACK, jackaudio.org), which is used to move music from one software package to another for audio, and the Advanced Linux Sound Architecture (ALSA, alsa-project.org) for MIDI.

If you are more into music composition, then Musescore (musescore.org) and Rosegarden (www.rosegardenmusic.com) may be more your speed. Musescore helps you make professional-looking sheet music, and Rosegarden is a music composition and editing environment based around a MIDI sequencer, which features a deep understanding of music notation but only includes basic support for digital audio. These free software tools are ideal for composers, musicians, and music students working in a small studio or home recording environments.

If you are serious about hosting your music, you can also set up your own streaming server with Sonerezh (www.sonerezh.bzh), which is a self-hosted web-based audio streaming application that allows access to your music from anywhere using your favorite web browser.

Making Musical Instruments

You can make and share your own physical musical instruments no matter what kind of music you are interested in. Musical instruments are normally broken down into five families: (1) strings, (2) percussion, (3) brass, (4) woodwinds,

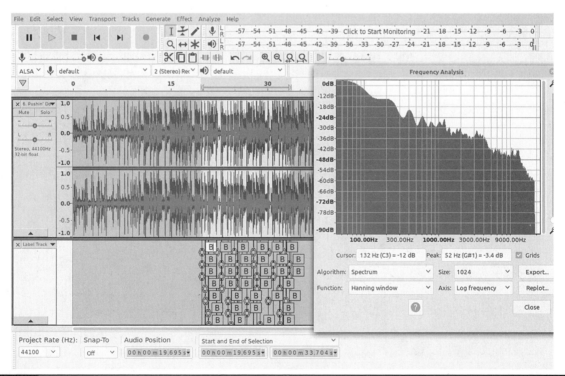

Figure 5.2   Screen capture of Audacity, an easy-to-use cross-platform open-source multitrack audio editor.

and (5) keyboards. The latter has largely been overtaken by the abilities of free and open-source software discussed earlier in "Open-Source Music Software." In addition, there are also commercial open-source music synthesizers such as the SparkFun SparkPunk Kit (www.sparkfun.com/products/11177), which is a sound generator/simple synthesizer made in the spirit of the Atari Punk Console. The SparkPunk has a ton of knobs and switches to provide a rich environment for tonal variations. For each of the other families of instruments, we will look at two examples of musical instruments whose designs are already freely available.

The Strings Family

If you want to make a string instrument, you have lots of choices. There is the hovalin (www.hovalabs.com/hova-instruments/hovalin), which is an approximately $70 functional acoustic violin that can be produced using open-source RepRap-class 3D printers. The hovalin's shape and dimensions (see the custom mashup sporting a double wolf's head instead of the traditional scroll in Figure 5.3) are inspired by the Stradivarius violin model, which sells for far more. A 3D-printed hovalin made of common plastic is not the quality of the original Stradivarius violins, but it is pretty good. You can also make low-volume hovalins for practicing without bothering those around you by printing in denser materials (e.g., the glycol-modified version of polyethylene terephthalate [PETG]). The hovalin project itself is a derivative of another open-source project, having been inspired by David Perry's FFFiddle (openfabpdx.com/modular-fiddle).

Figure 5.3 An open-source 3D-printed custom hovalin in play.

Perhaps guitars are more your thing. There is a dizzying array of open-source masterpieces to choose from, both acoustic and electric (all3dp.com/2/3d-printed-guitar-10-best-curated-models-to-3d-print). For example, Francesco Orrù's HR Giger Guitar is a tribute to Swiss painter H. R. Giger inspired by the Ibanez design (Figure 5.4).

Figure 5.4   Francesco Orrù's open-source 3D-printed HR Giger Guitar is a tribute to Swiss painter H. R. Giger. (CC BY-NC-SA) https://www.myminifactory.com/object/3d-print-hr-giger-guitar-6810

Websites like Instructables (instructables.com) cover how to make many stringed instruments from ancient lyres to mandolins and banjos out of pans. It is also possible to use open-source software to amp up your open hardware. Consider Guitarix (guitarix.org), a virtual guitar amplifier for Linux running on JACK (discussed earlier). Guitarix takes the signal from your guitar as any real amp would do, as a mono-signal input from your sound card, and then processes it by a main amp and a rack section. Both signals can be routed separately and deliver a processed stereo signal. You may fill the rack with effects from more than 25 built-in modules, including

stuff from a simple noise gate to brain-slashing modulation effects such as flanger, phaser, and auto wah, along with a long list of LV2 plugins. You can record your playing with an open-source recorder such as Ardour (ardour.org)—and, of course, edit and mix afterward. Guitarix was designed for guitar signals, but you can use it with other instruments or even a synthesizer (such as Qsynth [qsynth.sourceforge.io]) or any other sound generator.

The Percussion Family

Building percussion instruments is relatively straightforward compared with building other types of instruments. Many types of physical drums can be built from free, shared plans, from the least complex (turn over a 5-gallon paint can) to the complicated (turning a 55-gallon barrel of 17- or 18-gauge steel into a steal drum; www.steelpan-steeldrums-information.com/make-steel-pans.html). However, even percussion instruments can be automated with a little help from open-source electronics, such as the Arduino-powered robotic glockenspiel, called a Spielatron, as shown in Figure 5.5.

Figure 5.5   The Spielatron, an open-source robotic glockenspiel developed by Averton Engineering. (CC BY-NC-SA) https://www.instructables.com/id/Making-the-Spielatron-Robotic-Glockenspiel/

Drums can go completely electronic, like the Shapeshifter (gitlab.com/Faselunare/

shapeshifter), which is a free and open-source software–hardware combination expandable drum machine. Or you can go with the software-only Hydrogen (hydrogen-music.org), an open-source drum sampler software package.

The Brass Family

Making open-source brass family instruments is perhaps the most challenging. First, they are normally made out of metal (e.g., trumpets, trombones, French horns, etc.), and in general, working with metal is more challenging than working with other materials. Although there are open-source metal 3D printers, their resolution is still not at the level of the polymer printers discussed in Chapter 12 (Anzalone et al., 2013; Nilsiam et al., 2015; Nilsiam, Sanders, and Pearce, 2017). In addition, the tricky business of making metal 3D printers usable for average people (e.g., removing the welded substrate from the surface [Haselhuhn et al., 2014, 2015]) has only begun. The view of most scholars is that the term brass instrument should be defined by the way the sound is made, not by whether the instrument is actually made of brass. There is a 3D-printable trumpet published on Instructables (www.instructables.com/id/3D-Printable -Trumpet). It is made of plastic and not particularly well-developed yet. Fortunately, many "brass" instruments not made with brass, such as the didgeridoo, are easily made by hand. The didgeridoo was crafted in wood by indigenous aboriginal communities in Australia for thousands of years. Didgeridoos are made from hollowed cylindrical or conical tubes and use a beeswax mouthpiece. They are monster instruments that can be as tall as an adult and a few inches in diameter. In general, the longer the didgeridoo, the lower is the pitch or key of the instrument. An online bamboo dealer has excellent instructions on how to make a bamboo didgeridoo (www.guaduabamboo.com/working-with-bamboo/bamboo-didgeridoo) in about a

half hour after you have assembled the tools and supplies.

The Woodwinds Family

Unlike brass instruments, there are many open-source woodwind instruments. For example, you can select from easy directions for making an open-source flute out of wood from Cut The Wood (cutthewood.com/diy/how-to-make-a -flute-out-of-wood) or out of polyvinyl chloride (PVC) from Instructables (instructables.com/ id/Making-Simple-PVC-Flutes) or even out of metal pipes from Our Pastimes (ourpastimes.com/ how-to-make-a-piccolo-instrument-12599898 .html). Traditional flute designs are extremely well-documented (flutopedia.com/crafting.htm). Traditional flutes, such as the Native American Sparrow Hawk flute (www.thingiverse.com/ thing:2000241), can also be fabricated using these digital technologies.

In addition, there is an early open-source community centered around woodwinds made of brass. They are just getting started with the Open Source Saxophone Project (opensourcesaxophoneproject.com), whose goal is to spread knowledge and democratize the repair and eventual manufacture of saxophones. The Open Source Saxophone Project is wide-ranging, serving as a resource for saxophonists to become better consumers, providing a multimedia guide for repairers to improve their skills, and is moving toward the eventual goal of having the tools, computed-aided design (CAD) files, tutorials, and suppliers to enable anyone with the courage to build their own saxophone from scratch to do so.

Open Theremin

No matter what your musical taste, the open-source music community has you covered. Even if you are just looking for esoteric music for your open-source films (covered in Chapter 7), the open-source community has developed good options. For example, the theremin (shown in Figure 5.6) is not the most popular electronic musical instrument, but it is fun to play, even when you are controlling it without touching it. The Open Theremin (www.gaudi.ch/OpenTheremin) project helps you create those eerie sounds you hear in scary movies, so if you are making a movie, you can make your own sound track.

Figure 5.6 Screen capture of Coralie Ehinger playing the open theremin. http://www.gaudi.ch/OpenTheremin/index.php

References

Anzalone GC, Zhang C, Wijnen B, et al. 2013. A low-cost open-source metal 3D printer. *IEEE Access* 1:803–810.

Haselhuhn AS, Gooding EJ, Glover AG, et al. 2014. Substrate release mechanisms for gas metal arc weld 3D aluminum metal printing. *3D Printing and Additive Manufacturing* 1(4):204–209.

Haselhuhn AS, Wijnen B, Anzalone GC, et al. 2015. In situ formation of substrate release mechanisms for gas metal arc weld metal 3D printing. *Journal of Materials Processing Technology* 226:50–59.

Nilsiam Y, Haselhuhn A, Wijnen B, et al. 2015. Integrated voltage: Current monitoring and control of gas metal arc weld magnetic ball-jointed open source 3D printer. *Machines* 3(4):339–351.

Nilsiam Y, Sanders P, Pearce JM. 2017. Slicer and process improvements for open-source GMAW-based metal 3D printing. *Additive Manufacturing* 18:110–120.

CHAPTER 6

Scanning, Making Paper and Audio Books, and Sharing Books

Open-Source Book Scanning

The idea that all of human knowledge belongs to all of us is a difficult concept to actualize in a world dominated by copyrights. The concept of copyright, however, is meant to provide authors with an incentive based on a monopoly for a given amount of time, after which the copyrighted work enters the public domain for everyone to enjoy. This is straightforward.

Things get tricky, however, when laws change to benefit a few at the expense of the many. This is illustrated in the expansion of copyright law in the United States, shown in Figure 6.1. Much of the time, these changes effectively took works in the public domain and reverted them to the old copyright owners, often even after the original author's death. It is hard not to see the corruption in these changes in law. The 1998 Act extended the copyright of a poem you write

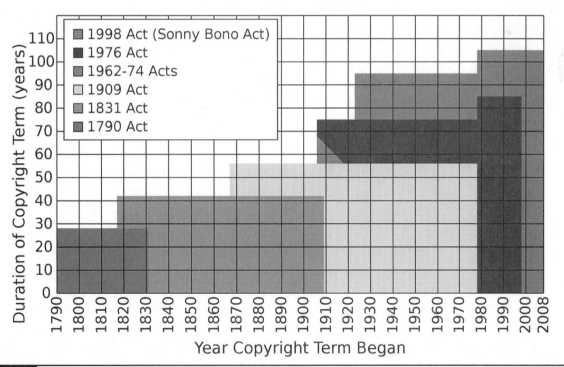

Figure 6.1 Expansion of U.S. copyright law (assuming that authors create their works 35 years before their death). Original author, Tom Bell, released this on his website. (CC BY-SA) https://en.wikipedia.org/wiki/Copyright_Term_Extension_Act

yourself for your lifetime plus 70 years. Seventy years after you are dead! But, if you are working for a corporation to which you signed over the copyright of your poem as part of the terms of your employment, then it lasts for 120 years after its creation or 95 years after its publication, whichever end is earlier.

This means that if the creator dies in the year of creation, the corporation has more rights than the creator (or his or her family). The same law holds for at least 25 years after creation and for any creative work that can be copyrighted, including music and sound recordings, websites, and printed works such as books. Why exactly should a corporation get rights to your work for 50 more years than you do if you made the work for yourself rather than as an employee?

Both the number of years and the disparity between individuals and corporations are pretty absurd, but proponents successfully argued in court three key justifications for the copyright extension: that public domain works will be (1) underutilized and less available, (2) will be oversaturated by poor-quality copies, and (3) poor-quality derivative works will harm the reputation of the original works. All three of these justifications were proven false in a detailed study that compared works from the two decades surrounding 1923 and made available as audiobooks (Buccafusco and Heald, 2013). This study found that copyrighted works were significantly less likely to be available than public-domain works, that there was no evidence of overexploitation driving down the prices of works, and perhaps most important, that the quality of the audiobook recordings did not significantly affect the prices people were willing to pay for the books in print (Buccafusco and Heald, 2013).

Despite these rule changes and corrupt laws, there is still a significant quantity of books in the public domain. These books can be legally digitized and shared with the world. If you have some of these old books, there are some open-source solutions, such as book scanners, to make the digitization easy. The cheapest way to make a book scanner is to use a single camera and a cardboard box (Reetz, 2019). Another super low-cost approach is to use a 3D-printable stand and a smartphone (RSID, 2018). However, many automated approaches have been developed by the do-it-yourself (DIY) book scanner community (www.diybookscanner.org) to preserve and share knowledge. You can buy a kit or build an automated book scanner from scratch from free plans.

For an industrial approach that does not hurt the book, consider the page-turning open-source Linear Book Scanner (linearbookscanner.org). Your book moves back and forth over the machine, and each time across, a vacuum sucks a page from one side to the other. As they are moving, the pages are scanned as they travel across two imaging sensors. The software that comes with the device produces a searchable PDF file of the book (e.g., providing added value over the original, which may have been printed long before computers were around).

Making Books

If you don't have the patience to wait for the copyright to run out, an easier approach to making good books available in the public domain is to write them yourself! There are a lot of open-source technologies that can help you do this. The book you are reading now was written primarily on a Systems 76 laptop running Linux Mint (linuxmint.com). The actual typing was done mostly in Libre Office (www.libreoffice.org), a free office suite. Libre Office is more than good enough for most books. If you want to get fancy, there is also Scribus (scribus.net), which is a page layout program, or you could do the entire book in LaTeX (www.latex-project.org), which is a high-quality typesetting system. You can also use all the open-source photo-editing and art programs discussed in Chapters 3 and 4.

Making Audio Books

Another way you can help the global community and have a little fun with a book in the public domain is to make it into an audio book. This is easy with most computers or smartphones. Just read the book and record. You can clean up your recordings with Audacity (sourceforge.net/projects/audacity), a free audio mixer. When you are done, you can post your full audio version on LibriVox (librivox.org), which hosts free public-domain audio books.

Reading and Sharing Books

If you want to read the books you have scanned (or really any other books), there is a host of free and open-source e-book reading programs, such as Calibre, Cool Reader, and FBReader. As one of my colleagues explained, "The day I discovered the open-source FBReader application was like Christmas. For someone who likes to read a lot of stuff but has to deal with the issue of limited storage space, e-pub files are the perfect solution. The FBReader allows me to read anything, everywhere, and anytime. How cool is that?" My colleague is not alone. The multiplatform e-book reader FBReader (fbreader.org) is used by more than 20 million people throughout the world. It supports all the popular e-book formats: EPUB, FB2, Mobi, Rich Text Format (RTF), Hypertext Markup Language (HTML), Plain Text, and many more. FBReader also provides direct access to popular network libraries that contain a large set of e-books. Download books for free or for a fee, and you can customize the whole experience. There are a lot of places to get free e-books (Table 6.1).

Table 6.1 Free e-Book Websites

Free Books Website Name	URL	Number of Available Books	Genres and Types of Books
Europeana	https://www.europeana.eu/portal/en	>50 mil	Artwork, artifacts, books, films, and music from European museums, galleries, libraries, and archives
Digital Public Library of America	http://dp.la/	>35 mil	Images, texts, videos, and sounds from across the United States
Internet Archive	http://archive.org/details/texts	>20 mil	E-books (mostly scholarly), many with DAISY files for print disabled people
The Online Books Page	http://digital.library.upenn.edu/books/	>3mil	E-books
Open Library	http://openlibrary.org/	>2.5 mil	E-books in a variety of subjects, many with DAISY files for print-disabled people
Google Book Search	http://books.google.com/	>1mil	Intends on scanning all existing books in the world, controls access to full text through copyrights
DailyLit	https://www.dailylit.com/	>100k	Emphasis on fiction
Project Gutenberg	https://www.gutenberg.org/	>60k	E-books in all the popular formats over a wide variety of subjects

(continued on next page)

Table 6.1 Free e-Book Websites (*continued*)

Free Books Website Name	URL	Number of Available Books	Genres and Types of Books
Manybooks	http://manybooks.net/	>50k	Free/public-domain e-books in many genres
LibriVox	https://librivox.org/	>13k	Audiobooks read by volunteers around the world
The Literature Network	http://www.online-literature.com/	>9k	Classic literature
Loyal Books (Books Should Be Free)	http://www.loyalbooks.com/	>7k	Draws from other free libraries to provide public-domain books
Wolne Lektury	http://wolnelektury.pl/	>5k	Polish e-books
Questia	https://www.questia.com/library/free-books	>5k	Classic and rare books
Feedbooks	http://www.feedbooks.com/	>4k	Public-domain books in many subjects
Open Culture	http://www.openculture.com/	>1k	Texbooks, e-books, and audiobooks
Classic Reader	http://www.classicreader.com/	>1k	E-books by classic authors
Online Library of Liberty	https://oll.libertyfund.org/	>1k	Overtly political library with niche collection of scholarly books about free market and liberty
Standard Ebooks	https://standardebooks.org/ebooks/	<1k	E-books focused on quality not quantity
LoudLit	http://loudlit.org/	<1k	Similar to Librivox
Projekti Lönnrot	http://www.lonnrot.net/	<1k	Finnish e-books
Authorama	http://www.authorama.com/	<1k	Brings public-domain textbooks into HTML
Legamus	http://legamus.eu/	<1k	Audiobooks
Classic Literature Library	http://classic-literature.co.uk/	<1k	Library of only classic works
Great Books and Classics	http://www.grtbooks.com/	<1k	Classic works by great philosophers
Planet Publish	http://www.planetpublish.com/	<1k	Free classic e-books in pdf format
Classical Chinese Literature	http://zhongwen.com/gudian.htm		Collection of very in-depth Chinese literature with learning tools for English speakers
World Public Library	http://worldlibrary.org/		Full database not free, but some classic works/poetry are
Christian Classics Etheral Library	http://www.ccel.org/	1.1k	Biblical commentary works
OReilly Open Books Project	https://www.oreilly.com/openbook/		Technical audiobooks (programming languages, etc.) great for learning software development

Project Gutenberg (www.gutenberg.org) is a good example of a free e-book website because it is a reasonable digital library with more than 60,000 free e-books that focuses on older works for which the U.S. copyright has expired. Project Gutenberg's volunteers digitized these works, and now the site allows you to either read the books online or download them to use on your favorite e-book reader as either a free e-pub or even a Kindle e-book. Project Gutenberg makes it easy to join the open-source book community. Because everything on Project Gutenberg is completely zero cost, the site asks for help to digitize more books (see "Open-Source Book Scanning" earlier in this chapter). You can join the distributed proofreading and formatting group, record audio books at LibreVox (librivox.org), or even simply report errors. What Project Gutenberg does not do is help you publish and share your own book. To be included on the site, you need to be published already.

Maybe you have some ideas for the next great novel or even a children's book, and maybe you are considering sharing it with the rest of the world with an open-source license. There are many on-demand book publishers now that will be more than happy to help you publish your book. Their printing costs are not overly expensive, particularly in bulk. However, it still gets expensive fast to make many hard copies of an open-source book available. It is much easier to share a digital version. There are also many websites where you can post a digital version of your book for free under an open license (e.g., at Freebooks [www.free-ebooks.net]) or target a specific genre, such as books for children (e.g., Free Kids Books [freekidsbooks.org]).

Cory Doctorow Makers

Cory Doctorow is the ex-director of the Electronic Frontier Foundation (www.eff.org), which is the leading nonprofit organization defending digital privacy, free speech, and innovation. Doctorow is also a science fiction author who both writes full time and allows you to download all his books for free (craphound. com). Doctorow released his first novel, *Down and Out in the Magic Kingdom*. under a Creative Commons license, which allows readers to download the book for free.

Now all his books come this way. How does he possibly do this? To give you an idea, the first Doctorow book I ever read was *Makers* (craphound.com/makers), an excellent fictional introduction to the subculture of makers and how America might still have a happy landing despite all the negative trajectories we are on. Since then, I have purchased more copies of this book than any other book I have ever read—to give as gifts to my students. Yet *Makers*, like all of Doctorow's other books, can be downloaded for free.

Doctorow explains his reasoning in three steps (Royal, 2013). He says that, realistically, it takes one or two clicks to download a book without having to pay for it anyway because illegal copies of most books are easily found. Therefore, by being generous and trusting his readers, he hopes that he can channel their energy into helping him. In other words, by sharing his book for free, he encourages more people to read it and thus more people will know about it and potentially buy a copy. He explains, "I'm not really concerned with being sure that everybody who reads my books pays for them. I'm much more concerned with making sure that everybody who's willing to pay gets a chance to read them." (Royal, 2013). Moreover, Doctorow feels that given today's technological landscape, modern art (such as books) will be copied, so he derives artistic satisfaction from allowing people to copy his books. Lastly, he finds this approach to be moral because ever-increasingly draconian measures are being enacted to defend copyrights in this internet era, to make it as hard as possible to prevent copying.

This, of course, is not working. As Doctorow explains, "More people copy more things now than they ever did, but they are monotonically increasing the amount of surveillance and censorship on the internet. And as an artist, I think that it's my duty not to have my works form part of the rationale for increasing censorship and control and surveillance on this amazing information medium that we all use for everything" (Royal, 2013). His solution is to avoid anything that leads you to demand censorship and surveillance on the internet.

Doctorow makes some powerful points (Doctorow, 2009). Most books hardly make any money and receive less readership than even the most benign, boring, and unimportant celebrity tweet. They will be available for free in minutes after publishing anyway, and yet most publishers are highly resistant to publishing open-access e-books. This puts authors in a challenging situation. Finding a publisher willing to publish a hard-copy book under a Creative Commons license is still pretty challenging.

Some publishers, particularly academic publishers, may be more receptive. My first book, *Open-Source Lab: How to Build Your Own Hardware and Reduce Research Costs*, was published by Elsevier, the largest scientific publisher. Because of the nature of the book, the company agreed to add a clause in the contract to make the book available for free, although it controlled how this would be done. What the company ended up doing was making one chapter at a time downloadable for free from its website each month. With patience, anyone could have the entire book for free completely legally.

Not surprisingly, however, shortly after the book was published, someone (or more than one person) uploaded a copy (or copies) to illegal download sites, and now there are free copies of it all over the place. Regardless, people still buy the book, and most important, I keep the information on that topic up to date on a CC BY-SA website (www.appropedia.org/Open-source_Lab). As discussed in Chapter 13, the academic community has turned the corner, so to speak. Open access in science writing and open-source software and hardware for science are becoming the norm.

This is not yet true for everything else, which brings us to the slightly uncomfortable topic of the copyright of the book you are reading now. Ideally, it would be open source as well, yet still published with the full support of a major publisher. For this book, I made the choice to partner with the publisher under a traditional contract because the publisher could provide the largest possible readership using conventional marketing. The core of this book—all the links and designs—is still open source and is available on the book's website (www.appropedia.org/Create).

References

Buccafusco C, Heald PJ. 2013. Do bad things happen when works enter the public domain? Empirical tests of copyright term extension. *Berkeley Technology Law Journal*, 1:1–43. https://ssrn.com/abstract=2130008

Doctorow C. 2009. Why free ebooks should be part of the plot for writers. *The Guardian*. https://www.theguardian.com/technology/2009/aug/18/free-ebooks-cory-doctorow

Reetz D. 2019. Bargain-price book scanner from a cardboard box. Instructables, https://www.instructables.com/id/Bargain-Price-Book-Scanner-From-A-Cardboard-Box/

Royal D. 2013. Author Cory Doctorow talks tech, Trotsky, what he's reading, and why he's giving away his books for free. *Phoenix New Times*. February 6, 2013. https://www.phoenixnewtimes.com/arts/author-cory-doctorow-talks-tech-trotsky-what-hes-reading-and-why-hes-giving-away-his-books-for-free-6570042

RSID. 2018. Document Book Scanner for CellPhone by siderits [WWW Document]. https://www.thingiverse.com/thing:3194536

CHAPTER 7

Making, Editing, and Sharing Videos

Open-Source Movie Cameras

Maybe the still images that we talked about in Chapter 2 are not enough for you, and you want to make movies. You are not alone—amazingly for people of my generation, most American youth would rather be a YouTuber than an astronaut (Berger, 2019). Anyone can shoot YouTube or Vimeo videos with a smartphone or a webcam, but you can also use open-source technology to get started with professional video cameras. Consider the AXIOM Beta (www.apertus.org), a professional digital cinema camera built around free and open-source software (FOSS) and open-hardware licenses. The AXIOM (Figure 7.1) comes with a sensor with 4K resolution, 35-mm diameter, and a global shutter with up to 300 frames per second at full resolution. With AXIOM footage, you can use Open Cine (www.apertus.org/opencine), which is open-source software that allows management, playback, and processing of the media captured by the camera.

Figure 7.1 Axiom open-source camera. (GPL v3) https://www.apertus.org/axiom-beta

Open-Source Video Editing

After you have your video, whether you used one of the advanced open-source cinematic cameras or simply used your smartphone, you will need to edit it. Fortunately, there are a number of open-source video editing programs.

Kdenlive (kdenlive.org) is a FOSS video editor that has a fully functional timeline that supports unlimited video/audio tracks, visible audio waveforms, preview rendering, and playback shortcuts. It comes with a large set of transitions, effects, and filters, and Kdenlive makes it simple to drag them onto clips, modify their settings, and see a live preview. When you're done editing and ready to export your finished video, you can choose from a large number of file types and presets.

Another popular open-source video editor is OpenShot (www.openshot.org) because of its simple and user-friendly interface, as shown in Figure 7.2. It comes with built-in tutorials when you first launch the software, and with the full user guide available on the website, OpenShot provides pretty easy video editing, even for beginners. You can drag and drop media into OpenShot to import them and then drag and drop them into the clips on the timeline. You can edit video and audio separately on an unlimited number of tracks. The program has a decent number of transitions and effects, as well as 3D animated titles if you also couple it with Blender.

If you are more adventurous, you can consider Blender (www.blender.org). Blender is far more than a video editor. I use it to teach students how to do 3D modeling and sculpting for my 3D-printing class. It can also be used for pure 3D graphics and 3D game development, painting, and making absurdly complex animations (Figure 7.3). Because it is so rich in features, it can be a little challenging to find the Blender Video Sequence Editor (VSE; *hint:* hold SHIFT + F8). The user community behind

Figure 7.2   The interface for OpenShot, a free and open-source video editor, showing a presentation involving massive collaboration efforts to build a barn in the Amish community.

Figure 7.3   Screenshot from *Spring*, a Blender open movie. (CC BY-4.0) https://cloud.blender.org/p/spring/

Blender, however, is really strong, so you can use either the manual (docs.blender.org/manual) or YouTube tutorials. Overall, Blender is a full-featured nonlinear editor with a multitrack timeline, cutting and trimming tools, and lots of keyboard shortcuts that (once you learn them) can make editing a breeze. For example, Blender has markers you can set in real time with the "m" key, as you are watching your clip, to indicate where you want cuts. After you're done watching the base video, you can easily skip to each marker, make a cut, and nondestructively delete the clips that you just cut. Then, Blender helps you automatically close the gap between each of the cuts and add transitions, if you need them.

Like the very best editors, Blender lets you unlink the audio and the video. This makes professional videos far easier to make. Of course, there is a plethora of more advanced options that allow you to turn your family videos into professional films. Then, of course, you can always add 3D graphics and animation if they're something you're into—or if the software inspires you to give it a shot.

What if you are a VJ (video jockey) and need to edit in real time for a performance? Open

source has an answer for that too with LiVES! LiVES (lives-video.com) is a nonlinear video editor that also has a number of real-time editing features that let VJs mix and control video clips to go along with audio—all on the fly at a live performance. LiVES has a clip editor, where you can apply effects such as fades, swirls, and colors to the videos you've imported, and then arrange the clips on the other part of the interface, the multitrack timeline, to render immediately or save for later. LiVES lets you create a custom key map that allows you to use effects or to transition between clips at the press of a button. This is super-helpful for VJing, but also makes any feature you use a lot easier to access and use. Of course, you can also "scratch" your videos backward and forward just like an old-school disk jockey on a vinyl record. There are many other more conventional open-source video editors such as Shotcut (shotcut.org), FlowBlade (jliljebl.github.io/flowblade/), Avidemux (avidemux.sourceforge.net), and Natron (natrongithub.github.io/, which does video effects [VFX]), so you're bound to find one that's just right for your video project.

Open-Source B-Roll and Sharing Your Videos

No matter what video editor you use, you will need content, and you may not have the time or resources to get it yourself. When you are making good open-source videos, you may want some free B-roll to enhance your own footage. Likewise, you may want to share your videos with others.

One of the best places to find these videos is the Internet Archive. The Internet Archive is a nonprofit library of millions of free books, software, music, websites, and more, including a lot of open-source movies (archive.org/details/opensource_movies). The movies are available under several types of licenses. Some licenses are accepted in the commons. For example, the Internet Archive houses the Prelinger Archives (archive.org/details/prelinger), which include more than 2,000 public-domain videos of historic significance. Other video sites use the Internet Archive this way. For example, Beachfront B-Roll (beachfrontprod.blogspot.com) is a blog offering free, royalty-free high-definition (HD) stock video, HD and high-dynamic-range (HDR) time-lapse clips, and HD animated looping backgrounds for any use. Download what you want by right-clicking and saving because the files are hosted by Archive.org.

There are also a lot of traditional websites dedicated to sharing open-source videos. For example, Pixabay Videos (pixabay.com/videos) provides copyright-free HD videos and footage released under Creative Commons CC0, and Flickr has a large assortment of Creative Commons videos (www.flickr.com). To find them, specifically search for Creative Commons. Similarly, Free Stock Footage Archive (freestockfootagearchive.com) provides videos under a CC BY-3.0 license, as does Videvo Stock Footage (www.videvo.net), which provides royalty-free stock video footage under a CC

BY-3.0 license or licensed by the company for easy use in noncommercial and commercial projects.

Various governments also provide free video. For example, European Southern Observatory videos covering space and astronomy are licensed under Creative Commons Attribution CC BY (www.eso.org/gallery/v/Videos), as are NASA videos (www.nasa.gov/multimedia/videogallery). Similarly, the Netherlands Institute for Sound and Vision houses the Open Images Project (www.openbeelden.nl). Because many of the videos contain Dutch audio, it is helpful to use one of the editors from the preceding section that can separate the audio and video. Not to be outdone, the official website of the president of the Russian Federation (kremlin.ru) provides all materials (e.g., photos, texts, videos, audios, etc.) licensed under CC BY-3.0.

There are also many more mainstream video sharing sites that enable you to post your own work with an open-source license. YouTube (www.youtube.com) is the biggest and best known; it enables you to share your video and get it seen by a large audience. The default YouTube license is called the *Standard YouTube License*, which means that you grant the broadcasting rights to YouTube (i.e., Google). With this license, your video can only be accessed from YouTube and cannot be reproduced or distributed in any other form without your consent. You give your video to YouTube to use for generating advertising revenue (for those not using open-source ad blockers, we discuss this in Chapter 2). You can, however, choose an open-source license by selecting the "Creative Commons" option. With Creative Commons licenses, you grant permission for everyone to use your video, to edit it, and to redistribute it (even commercially). To find videos on YouTube with a Creative Commons license, first search for what you want video footage of and then click on the filter and select "Creative Commons" from the feature choice.

Similarly, Vimeo (vimeo.com) allows members to enable Creative Commons licenses that grant copyright permissions on their videos so that others can copy, distribute, edit, remix, and build on them while giving credit to the original video's creator. A full list of the Creative Commons licensed videos on Vimeo can be found at vimeo.com/creativecommons. As of this writing, millions of Creative Commons videos are available, each broken down into the specific Creative Commons license.

Reference

Berger E. 2019. American kids would much rather be YouTubers than astronauts. *Ars Technica*. https://arstechnica.com/science/2019/07/american-kids-would-much-rather-be-youtubers-than-astronauts/

CHAPTER 8

Making and Sharing Maps and GIS Data

GIS stands for *Geographic Information System*, which is a system designed to capture, store, manipulate, analyze, manage, and present spatial or geographic data. That is a mouthful—but basically GIS lets you go much deeper into spatial and geographic knowledge than a simple map ever could. If you have ever used a Garmin or your smartphone to get driving directions, you are already familiar with how useful GIS technology can be. In this chapter, we will look at how to use open-source GIS and maps to bring some geographic firepower to your projects.

Open-Source Maps

Maps are useful for planning a trip or just gathering information about our world. Many of us depend on maps for getting around, which is why it is nice to have access to free and open-source maps with no strings attached. For basic images of world maps or maps of cities, MapsOpenSource (mapsopensource.com) provides geographic menus as well as a search feature for finding them. For a street map you can actually use for navigating the world, consider OpenStreetMap (www.openstreetmap.org), which is created by people like you and is free to use under an open license (Figure 8.1). It is truly a crowd-sourced mapping project

that is built by a community of mappers who contribute and maintain data about a long list of useful features that focus on local knowledge all over the world, including roads, trails, cafés, railway stations, and other attractions. You can help to improve OpenStreetMap with high-tech projects such as using aerial imagery from open-source drones (see the section "Open-Source Drones" in this chapter), balloons, or kites. After pictures are taken, OpenDroneMap (opendronemap.github.io), a free and open-source software, allows you to postprocess drone, balloon, kite, and street-view data to geographic data, including orthophotos, point clouds, and textured mesh.

Likewise, open-source Global Positioning System (GPS) devices (see the section "Open-Source GPS" in this chapter) can be used to improve the maps, which are particularly helpful for areas that do not photograph well (e.g., forested hiking, mountain biking, and cross-country ski trails). A lot of OpenStreetMap's usefulness, however, comes from contributors using low-tech field maps to verify that OpenStreetMap is accurate and up-to-date. You can contribute no matter how technically sophisticated you are, and your contribution will really matter. As with many of the other projects in this book, the open-source maps are useful on their own, but it is also gratifying to actually participate in making them better.

(a)

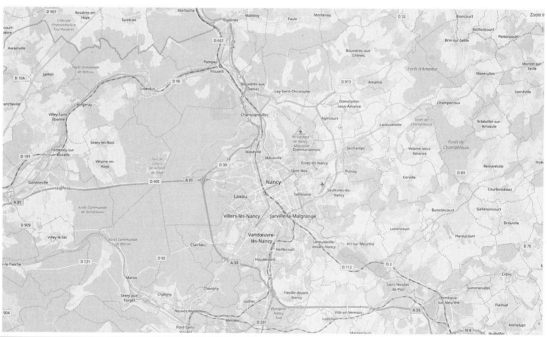

(b)

Figure 8.1 Screen captures of open street map zooming from (a) Europe to (b) Nancy, France, to (c) Place Stanislas with local restaurants labeled at full zoom. (CC BY-SA) https://www.openstreetmap.org/

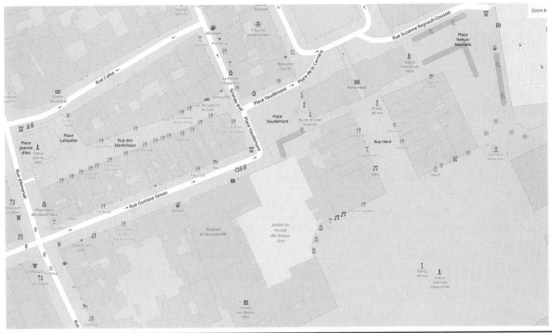

(c)

Figure 8.1   Screen captures of open street map zooming from (a) Europe to (b) Nancy, France, to (c) Place Stanislas with local restaurants labeled at full zoom. (CC BY-SA) https://www.openstreetmap.org/

Open-Source Drones

In many ways, open source has driven and led innovation in the drone technological space (Anderson, 2014). The primary innovation hub for drones is DIY Drones (diydrones.com), which is the largest community for amateur unmanned aerial vehicles (UAVs). DIY Drones created the community that built the ArduPilot (ardupilot.org), which is the world's first universal autopilot platform covering the air with planes, multirotors/multicopters, and helicopters, as well as the new-fangled quad planes and compound helicopters; the ground with many types of ground rovers; and the water with boats and even submarines. ArduPilot is not a toy—it is fully loaded and reliable open-source autopilot software that has been installed in more than a million vehicles worldwide. We know it is reliable because of its advanced data-logging, analysis, and simulation tools, making it easy to show that ArduPilot has been thoroughly tested and proven to be reliable

autopilot software. As with the other projects discussed in this book, the open-source nature of ArduPilot's code base means it is rapidly evolving, and always at the cutting edge of drone technology development. It is mature enough that ArduPilot has become its own platform (similar to the Arduino open-source electronics platform we discuss in Chapter 11 along with the origin of the ArduPilot name and technology). Today, many commercial companies (e.g., 3DR, jDrones, PrecisionHawk, AgEagle, and Kespry), several large institutions and corporations such as NASA, Intel, and Boeing, and countless colleges and universities around the world, build their own offerings from the ArduPilot code base and provide peripheral supplies and new interfaces. Thus, users benefit from a broad ecosystem of sensors, companion computers, and communications systems all constantly forming around the ArduPilot core. ArduPilot-controlled drones such as the octocopter shown in Figure 8.2 can be used for everything from search-and-rescue operations, agricultural

Figure 8.2 Autonomous coaxial octocopter flying with ArduPilot. (Image courtesy of Olivier J. Brousse.) (CC BY-SA-4.0) http://www.ardupilot.org/

Figure 8.3 Flone 1.0, available on Instructables, developed by Aeracoop: Aerial Cooperation. (CC BY-SA) https://www.instructables.com/id/Flone/

automation for robot tractors, underwater exploration of coral reefs, and inspection of roads, bridges, or oil pipelines, as well as, of course, aerial mapping.

On the hardware side, Pixhawk (pixhawk.org) is an independent open-hardware project that aims to provide readily available high-quality and low-cost autopilot hardware designs that can run ArduPilot, as well as PX4. PX4 (px4.io) is a pro-quality copter, plane, rover, and vertical takeoff and landing (VTOL) software stack from the Dronecode Foundation. The Dronecode Project, hosted under the Linux Foundation, is a vendor-neutral site for end-to-end drone software (www.dronecode.org).

If you want a lighter drone project and you love your smartphone, you might want to join Spain's Flone Project (flone.cc/en/home-2), which converts your smartphone into a drone while using another smartphone as the controller. It combines a digitally fabricated airframe (following what we cover in detail in Chapter 12) with free and open-source software that allows an open-source operating system (i.e., Android) smartphone on the ground to control the one strapped onto the airframe via Bluetooth (Figure 8.3).

There are many other open-source drone projects that enable open-source control over a much wider range of drone types. For example, Paparazzi UAV (Unmanned Aerial Vehicle, wiki.paparazziuav.org) is an open-source drone hardware and software project encompassing autopilot systems and ground station software for a long list of multicopters or multirotors, fixed-wing aircraft, helicopters, and hybrid aircraft. The main goal of Paparazzi UAV is to enable autonomous flight (even of multiple aircraft), although it is capable of manual flying. Paparazzi works as a dynamic flight plan system that uses way points as "variables." This makes it easy to create very complex, fully automated missions without the operators or users intervening in the control of the drone.

Finally, one of the key selling points for using open-source software and hardware for this particular technology, which has a wide range of applications, including military and law enforcement, is that because the source code is open, it can be audited to ensure compliance with security and secrecy requirements. This is particularly critical for sensitive hardware, which we discuss in Chapter 11.

Open-Source GPS

With GPS modules becoming a standard offering in many modern smartphones, the costs of GPS basic components have come down so far that they are easily accessible to makers. This provides the potential for many wondrous creations you can build on for your own projects—and the maker community has not disappointed! We discuss open-source electronics and the Arduino microcontroller in Chapter 11, but just on the Arduino GPS page, there are 38 open-source GPS projects (create.arduino.cc/ projecthub/projects/tags/gps) you might want to try out. There is, of course, navigation for cars and drones of all kinds, as you might expect, but makers have also used this open-source GPS technology to do more sophisticated tasks such as making accident and pothole detectors, suntracker controls for solar photovoltaic arrays, and devices to help and track kayak navigation.

Having the ability to use GPS any way they want allows people to just have fun—as with the autonomous "follow me" cooler, which is a small robot that carries around a cooler and dutifully follows a person holding a smartphone via Bluetooth. It uses GPS to navigate. Open-source GPS use is not limited to Arduino; for example, Raspberry Pi GPS projects include navigating with Navit or OpenMapTiles (openmaptiles. com), which provides world maps you can deploy on your own (e.g., the OpenStreetMaps discussed earlier in "Open-Source Maps").

Open-source GPS can help you with the oddest of modern problems. Does the rapidly growing Flat Earth Society count one of your friends or relatives as a member? Are you having trouble countering arguments they find on YouTube? Perhaps a backyard experiment will help. Combining a Raspberry Pi, data logger, GPS tracker, and camera with a helium balloon, you can make a fun project to show that the Earth is not flat. Handy open-source software

Figure 8.4 Earth picture as taken from a Raspberry Pi ballooning experiment. (Photo courtesy of the Raspberry Pi Foundation.) (CC BY-SA-4.0) https://opensource.com/ life/15/9/pi-sky-high-altitude-ballooning -raspberry-pi

can show the flight of your balloon in three dimensions on Google Earth and send back photographic evidence the whole way. It can also be used for real science or just to take your own "almost astronaut" pictures, as shown in Figure 8.4.

In addition to open hardware for GPS, there is also well-developed software with the Traccar Project (www.traccar.org), which is flexible enough to meet your needs no matter which end of the spectrum your own GPS project or GPS device resides. You can self-host the server or let it reside in the cloud. Traccar has a fully featured web interface that is amenable to a desktop or a mobile device, where it can be used as a GPS tracker in real time—mapped on roads or on satellite images. You can set up web notifications, along with support for email and short message service (SMS), which allows for external alerting in cases of fuel and maintenance events, geofencing (e.g., Do you want to make sure your teenager keeps the car in town?), and many other applications. Traccar supports simple location history, trip, chart, and summary reports, which you can use to track everything from a walk in the neighborhood to an epic cross-country road trip in the family RV. You can see this all through the web or mobile app or download the data to play with at home.

There are many other open-source GPS software solutions that can be scaled to any project. For example, if you are looking for something a little simpler, consider TrackIt! (github.com/flespi-software/TrackIt), which is a lightweight open-source solution for road navigation, as well as tracking your stuff with a simple intuitive interface and basic functionality. By contrast, if your fleet of open-source drones is growing, OpenGTS (opengts.sourceforge.net) is an open-source project designed specifically to provide web-based GPS tracking services for a fleet of vehicles.

GPS may not even be necessary for global location because a team of students from the Royal College of Art and Imperial College have designed an open-source alternative to GPS called *Aweigh* (www.aweigh.io; shown in Figure 8.5). Aweigh, as in "Anchors aweigh!", is a particularly clever device that outputs your latitude and longitude. Aweigh does

this by calculating the device's position using the sun—a feature inspired by the polarized vision of insects—and thus does not rely on satellites. Much like the Viking navigational aids known as *sun stones*, Aweigh can even work on cloudy days when the sun is obscured. Perhaps more important, unlike GPS devices that rely on satellites, Aweigh functions offline, so a user's positional data are held privately. This is particularly important given what has been disclosed by *The New York Times* on mass-scale smartphone tracking (Thompson and Warzel, 2019). Aweigh is open source, can be manufactured by open-source digital technologies (see Chapter 12), and uses a Raspberry Pi (see Chapter 11) as the computational core.

Open-Source GIS Information

There are many sources of open-source GIS data that you can use for your GIS projects. Some of the top sources are summarized in Table 8.1.

Open-Source GIS Software

GRASS GIS

So you have some GIS data and now you want to do something with it. Open source can come to the rescue again. There are a dozen good open-source GIS software tools. Here, we will look at two: GRASS and QGIS.

GRASS (Geographic Resources Analysis Support System) GIS (grass.osgeo.org) is a free and open-source GIS software suite used for geospatial data management and analysis, image processing, graphics and maps production, spatial modeling, and visualization. GRASS GIS is currently used in academic and commercial settings around the world, as well as by many governmental agencies and environmental

Figure 8.5 One of the many designs of the Aweigh open-source navigational aid created by Flora Weil, Keren Zhang, Samuel Iliffe, and States Lee. (CC BY) https://github .com/build-aweigh/build-aweigh/blob/ master/Logbook/Aweigh%20book.pdf

Table 8.1   Open-Source GIS Data Repositories

Open-Source GIS Repository (URL)	What Does It Have?	Licenses
ESRI Open Data (opendata.arcgis.com)	Free data sets from thousands of organizations; various download formats available; works best with ESRI data sets	Data-set-specific licensing
FAO GeoNetwork (www.fao.org/geonetwork)	Satellite imagery and spatial data specific to food and agriculture	GNU GPL and CC BY-3.0 unported licenses
ISCGM Global Map (globalmaps.github.io)	Archived global map data by region; specific strengths in land cover and tree cover	Free for noncommercial use
NASA Earth Observations (NEO) (neo.sci.gsfc.nasa.gov)	Large sets of satellite data in various formats (JPEG, PNG, Google Earth, and GeoTIFF); updated frequently	CC BY equivalent
NASA's Socioeconomic Data and Applications Center (SEDAC) (sedac.ciesin.columbia.edu)	Human interactions with the environment; socioeconomic data with thematic maps	CC BY equivalent
Natural Earth Data (www.naturalearthdata.com/downloads)	Key cultural and physical vector GIS data sets at a global scale; raster data sets for relief maps	Public domain
OpenStreetMap (gisgeography.com/openstreetmap-download-osm-data)	Crowdsourced, detailed data, but accuracy varies; cultural vector data.	Open Data Commons Open Database License
Open Topography (www.opentopography.org)	Provides LiDAR data (a less common data type with many uses) by region (90% North America)	CC BY equivalent
Sentinel Satellite Data (scihub.copernicus.eu/dhus)	Highest-resolution public satellite data; ortho basemaps and synthetic aperture radar data	"EU law grants free access to Copernicus Sentinel Data and Service Information for the purpose of the following use in so far as it is lawful: (a) reproduction; (b) distribution; (c) communication to the public; (d) adaptation, modification and combination with other data and information; (e) any combination of points (a) to (d)" (https://sentinels.copernicus.eu/documents/247904/690755/Sentinel_Data_Legal_Notice)
Terra Populus (www.terrapop.org)	Combines census data and environmental data from over 80 countries from 1960 onward to illustrate temporal and spatial changes over time	Cite the source of the data when used, but do not redistribute the data (https://terra.ipums.org/citation).
UNEP Environmental Data Explorer (geodata.grid.unep.ch)	Provides a large variety of spatial and nonspatial data	Data-set-specific licensing
U.S. Geological Service Earth Explorer (earthexplorer.usgs.gov)	Key for use finding remote sensing data; many search tools (filters, by region, etc.); data outside the United States available as well	Public domain

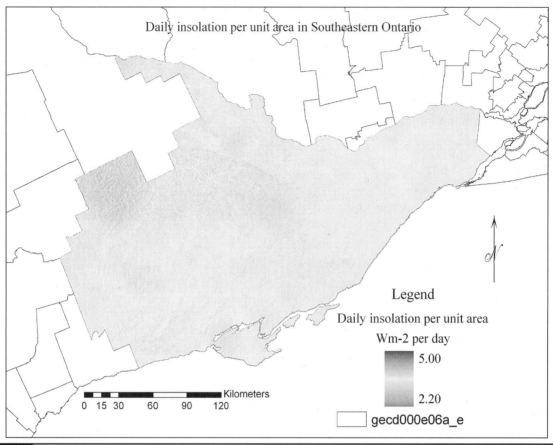

Figure 8.6   Using GRASS GIS and r.sun to map the daily insolation per unit area in southeastern Ontario. (Map provided by Ha Thanh Nguyen.) (CC BY-SA) https://www.appropedia.org/File:Daily_insolation_per_unit_area_in_Southeastern_Ontario.jpg

consulting companies. For example, my collaborator and I used it to help renewable-energy companies find some of the best places to build solar farms in southern Ontario (Nguyen and Pearce, 2010), as shown in Figure 8.6.

GRASS GIS is a founding member of the Open Source Geospatial Foundation (OSGeo). OSGeo (www.osgeo.org) houses a multitude of projects that will be useful for open-source GIS work of any kind, including content-management systems such as GeoNode, metadata catalogs, desktop programs such as GRASS GIS and QGIS, web-mapping programs such as MapFish and GeoMoose, geospatial libraires such as GeoTools, and spatial databases such as PostGIS.

QGIS

Another OSGeo project is QGIS (www.qgis.org). QGIS has a large open-source developer community and thus provides a continuously and rapidly expanding number of capabilities. You can visualize, manage, edit, and analyze data and compose printable maps. Core plugins include:

- Coordinate Capture (captures mouse coordinates in different coordinate reference systems)
- DB Manager (exchange, edit, and view layers and tables from or to databases and execute SQL queries)
- eVIS (visualize events)

- Geometry Checker (check geometries for errors)

- Georeferencer GDAL (add projection information to rasters using the Geospatial Data Abstraction Library)

- GPS Tools (load and import GPS data developed in Section 8.3)

- MetaSearch Catalogue Client (interacting with metadata catalog services supporting the OGC Catalog Service for the Web (CSW) standard)

- Offline Editing (allows offline editing and synchronizing with databases)

- Processing (the spatial data-processing framework for QGIS)

- Topology Checker (finds topological errors in vector layers)

And perhaps most interesting, these open-source packages play nicely together because there is a plugin GRASS 7 that integrates GRASS GIS into QGIS. QGIS can also be scripted with open-source Python. QGIS users share some of their work on Flickr (www.flickr.com/groups/qgis-screenshots), as can be seen in Figure 8.7, which shows Basa de la Mora (Ibón de Plan) lake in the Spanish Pyrenees.

With the open-source tools discussed in this chapter, the world is yours, so go enjoy it!

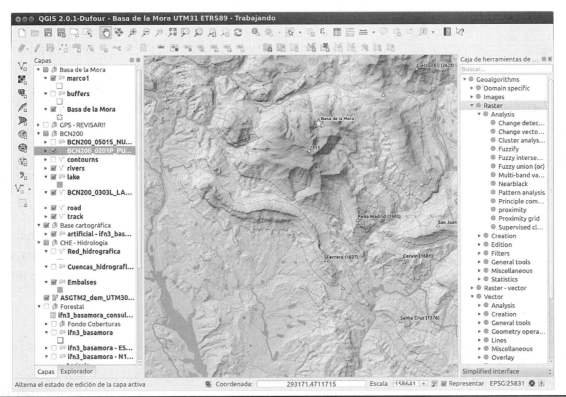

Figure 8.7 QGIS 2.0.1 running on GNU/Linux (Ubuntu 12.04) screenshot of a topographic map around Basa de la Mora (Ibón de Plan) lake, Spanish Pyrenees, by Miguel Sevilla-Callejo. (CC BY-SA-2.0) https://www.flickr.com/photos/msevilla/10288258133/in/pool-qgis-screenshots/

References

Anderson C. 2014. *Makers: The New Industrial Revolution*. Random House, New York.

Nguyen HT, Pearce JM. 2010. Estimating potential photovoltaic yield with r.sun and the open source Geographical Resources Analysis Support System. *Solar Energy* 84(5):831–843.

Thompson SA, Warzel C. 2019. Twelve million phones, one dataset, zero privacy. *The New York Times*.12-19-2019. https://www.nytimes.com/interactive/2019/12/19/opinion/location-tracking-cell-phone.html

CHAPTER 9

Making and Sharing Clothing

Much like the other areas of free and open-source design we have discussed, embracing open-source clothing provides you with far more options that you are accustomed to as a conventional consumer. Because of some vagaries of intellectual property law, the designs of clothing cannot be protected. In the fashion industry, innovation is fast because there is a general lack of artificial intellectual monopoly (Boldrin and Levine, 2008), and it is relatively easy to alter fashion design (Raustiala and Sprigman, 2012). This helps to explain why fashion is constantly changing styles (i.e., clothing companies cannot rely on 20-year patent protection, so they must innovate to keep selling products). When you look online for clothes or go to a major department store, it may at first appear that there is a lot of selection. At closer inspection, however, there are not that many options, and in a town with a few clothing stores, it is not uncommon for many people to be wearing the same thing—mass-produced items whose only customization is small, medium, and large.

Even well-known celebrities such as actress and singer Vanessa Hudgens and model and philanthropist Petra Nemcova battled it out over the same gold dress (Figure 9.1). The wealthy and famous don't seem to be able to escape the relatively limited selection created by the mass-produced clothing industry, which explains the popularity of the media meme

"Who wore it better?" The average person has very little clothing that is made to fit them exactly—perhaps a suit. With open-source clothing, everything you wear can be customized and made to fit you exactly. This provides a

Figure 9.1 Who wore it better? Where Petra Nemcova and Vanessa Hudgens battle it out over the same gold dress. Hudgens is a famous actress and singer, and Nemcova is a model and the cofounder and vice chair of All Hands And Hearts (AHAH)—Smart Response, which is a nonprofit dedicated to providing fast and effective support to communities hit by natural disasters worldwide. Live & Let Die </3. (CC BY-2.0) www.flickr.com/photos/shout-it/4104493149

new creative and positive means to express individuality.

Free Patterns for Clothes

Many websites now offer free patterns for pretty much any type of person or style. For example, FreeSewing (freesewing.org) is an open-source platform for made-to-measure sewing patterns—think of it as the "Wikipedia of sewing patterns in its early days." FreeSewing helps those who would like to make their own or loved one's clothing using the combined knowledge of the sewing community, its patterns, and its complete documentation. FreeSewing gives all of its core designs names, such as the "Wahid waistcoat" shown in Figure 9.2a. The designs are rated by difficulty, the number of measurements that you need to take to custom design it, and various options. If the Wahid is a bit too much for you to get started on because of the complexity, there are simpler designs, such as "Florent Flat cap" that only needs a single measurement (Figure 9.2b).

Other, more conventional pattern houses, such as BurdaStyle, have been captivated by the open-source philosophy of sharing and allowing the public to adapt its clothing to each individual's specific needs, so the company removed the copyright from its patterns (Style, 2009). The company's open-source sewing patterns are free to be used as a base for your own design, which you can wear, gift, or sell. Whatever you sew, you can sell it if you like. Sew Daily has even started selling how-to videos using BurdaStyle designs (Figure 9.3). BurdaStyle hopes that open sourcing its designs will inspire creativity and spawn multiple new designs. Even high-end

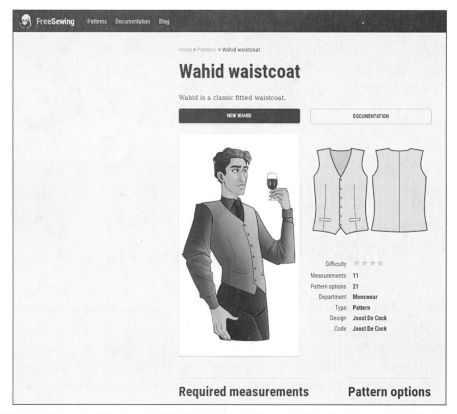

(a)

Figure 9.2   (a) Screenshot of Wahid waistcoat. FreeSewing. (Open-source MIT license) https://en.freesewing.org/

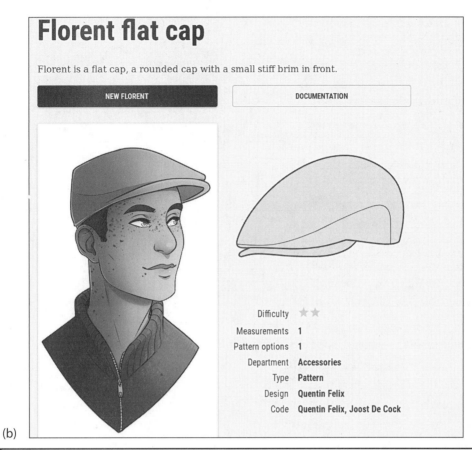

(b)

Figure 9.2 (b) Florent flat cap. FreeSewing. (Open-source MIT license) https://github.com/freesewing/freesewing

Figure 9.3 Screenshot of Sewdaily TV video *How to Sew the BurdaStyle Short Jumpsuit* (https://videos.sewdaily .com/courses/how-to-sew-the-burdastyle-short-jumpsuit). Open-source pattern can be downloaded from www.burdastyle.com/pattern_store/patterns/short-jumpsuit-052013

pattern companies such as Portia have gone open source. The company recognized that obscurity is a far larger risk than piracy. The company wants people to use and innovate its designs, and it only asks for attribution (Masnick, 2007). Many other firms behave similarly, providing free basic sewing patterns (www.moodfabrics.com/blog/category/basic-sewing-patterns and www.allfreesewing.com).

Making your own clothes also enables you to be a good environmental steward. You can repair clothes rather than discarding them in landfills. You can even use your old clothes to make new ones. In addition, one of the lowest-cost ways of getting fabric is to buy clothes on sale at secondhand stores and then disassemble them and make what you want from free patterns. Extralarge (XL) and bigger sizes are particularly good for this approach to the lowest cost per square meter of fabric. All these methods of up-cycling and reusing are good for the environment, as well as your pocketbook.

Open-source design does not stop with the clothing you wear on your back. It has also gone all the way down to your toes to create some truly one-of-a-kind shoe designs. A Fablab engineer in France designed the *chaussures à talons* (or "high heels shoes") that can be digitally manufactured, and released them under a GNU GPL license (Figure 9.4). Thus, they are free for anyone to make, and more than 1,000 people have already made a pair. If you are into high heels and this is not specialized enough for you, the open-source community has gotten truly original. For example, inspired by the TV show *Mr. Robot*, Naomi Wu designed an attractive pair of platform heels that can also be used for information security penetration testing (e.g., spying or probing of security threats), as shown in Figure 9.5a. As can be seen in Figure 9.5b, in one shoe, there is a USB keystroke

Figure 9.4   Chaussures à talons—high-heel shoes by experience. (GNU FDL) www.thingiverse.com/thing:915618

recorder, which is a pass-through device that goes into the back of the computer where you normally plug the keyboard in and records everything typed on the keyboard. This enables the collection of passwords, encrypted messages before encryption, and so on. Because Wu is an open-source hardware trend setter, the shoes also house a retractable Ethernet cable for the OpenWRT router (openwrt.org). Finally, no spy shoes would be complete without a full lock pick set housed in the other shoe.

Harnessing the open-source philosophy for shoes is not only done by makers. Following in the footsteps of free and open-source software (FOSS), Open Source Footwear gives everyone a chance to have a say in the shoes they want to see (www.fluevog.com/community/open-source-footwear). John Fluevog Shoes is a company that has essentially crowdsourced its designing work. If you have an idea for a new pair of shoes and want to see it in real life, you can submit your design for all to use. If the company likes your design, it will make you a pair and start to sell the shoes.

(a)

(b)

Figure 9.5 (a) Naomi Wu modeling her Wu Ying Shoes—Pentesting Platform Shoes; (b) the security-penetration testing tools that the shoes hide. (CC BY-SA) www.thingiverse.com/thing:980191

Open-Source Software for Automated Sewing

Ink Stitch (inkstitch.org) is an open-source machine embroidery design platform based on Inkscape (the vector graphics program we looked at in Chapter 4). You make or create a vector image in Inkscape, then covert it to embroidery vectors, plan the stitch order and attach commands, take a look at it with a built-

in simulator, and finally, try it out in real life. Then people can share these designs too, like Silver Seams (silverseams.com/free-patterns/embroidery-patterns). In addition, using 3D-printed parts, you can convert any normal sewing machine into a computer numerical control (CNC) embroidery machine (inkstitch.org/tutorials/embroidery-machine). These same concepts are used in the other automated digital tools discussed in Chapter 12. Similarly, Seamly (seamly.net) and Valentina (valentinaproject.bitbucket.io) are open-source pattern design software, which are designed to be the foundation of a new stack of open-source tools to remake the garment industry. The Valentina project has a strong social mission because those behind it believe that small-batch and custom-sized clothing manufacturing is essential to

1. Create a sustainable future,
2. Preserve small- to medium-sized textile spinning and weaving manufacturers,
3. Enable independent and small designers and manufacturers to scale up to make a decent living,
4. Rebuild local garment districts, and last but not least,
5. Reduce or eliminate slave labor.

Open Hardware for Clothes Making

Although open-source ecology has called for an open-source sewing machine, little progress has been made on that front to date. Open-source knitting, however, is much further along. OpenKnit (openknit.org) is a low-cost open-source digital fabrication tool that lets you create your own custom-knitted clothing from digital files. OpenKnit takes you from yarn straight to its end-use clothing (e.g., a sweater, as shown in Figure 9.6). This is done with almost no waste, similar to a 3D-printed object, because the product is knitted to shape. OpenKnit enables the design of digital custom clothing and production of the clothes wherever you happen to be, providing a lot of opportunity to flex your creativity. In turn, you can use others' digital designs or share your own at Do knIt Yourself (doknityourself.com), which acts as an open-source clothing platform to share clothes virtually and freely for anyone using OpenKnit machines and Knitic software (github.com/mcanet/knitic).

Figure 9.6 Screenshot of OpenKnit video fabricating a "Hello world" sweater. Openknit, a Reprap-inspired open-source knitting machine, 2014. Boing Boing. (BY NC-SA-3.0) https://boingboing.net/2014/02/20/openknit-a-reprap-inspired-op.html

Do-It-Yourself (DIY) Clothing and the Future

In the past, everyone made their own clothing. It was a point of pride and a way to demonstrate your skills. Perhaps we are headed back there

with both old-school and high-tech methods of making custom perfect-fit clothes for ourselves. The COVID-19 pandemic has certainly made the value of distributed manufacturing (Pearce, 2020), and of sewing in particular, clear. Throughout the world, people are using open-source patterns for face masks (e.g., www.craftpassion.com/face-mask-sewing-pattern) to make masks for their family, friends, and local hospitals to use in order to conserve N95 masks.

References

Boldrin M, Levine DK. 2008. *Against Intellectual Monopoly*. Cambridge University Press, Cambridge, UK.

Masnick M. 2007. Open Source . . . Sewing? *Techdirt*. https://www.techdirt.com/articles/20070413/110343.shtml

Pearce JM. 2020. Distributed manufacturing of open source medical hardware for pandemics. *Journal of Manufacturing and Materials Processing*, *4*(2), p. 49. https://doi.org/10.3390/jmmp4020049

Raustiala K, Sprigman C. 2012. *The Knockoff Economy: How Imitation Sparks Innovation*. Oxford University Press, Oxford, UK.

Style B. 2009. What Is Open Source Sewing? *BurdaStyle*. https://www.burdastyle.com/discussions/getting-started/topics/what-is-open-source-sewing

CHAPTER 10

Making and Sharing Woodworking and Other Old-School Skills

People who work on old-school skills such as woodworking or blacksmithing are normally more than happy to help the next generation learn. For example, Christopher Schwarz (2014), at Lost Art Press, freely allows anyone to use his writings and designs for woodworking classes, presentations, or to make products to sell (e.g., workbenches or chests). As one of the leading authors in the field, he points out: "To be honest, little that I do (or any other woodworking author, for that matter) is original. My work is inspired by old books, new books, old work and new. If my work inspires you to teach others about it, I'm happy. If it makes you want to share it with your club, ditto. And if it helps you make something you can sell, that's great, too" (Schwarz, 2014). This type of generous sharing makes it easy to pick up skills if you happen to live near an appropriate makerspace, hackerspace, or fablab (e.g., MakerMaps in Figure 1.5). Many of these places have classical tools, such as those needed for traditional woodworking. Even without access to a physical community, knowledge-sharing sites can provide woodworking plans and how-to videos, shared by great people like Schwarz.

Some of this sharing has gone online only and has become a true open-source furniture movement. These projects can be pure do-it-yourself (DIY) builds, such as the wall-mounted combination work desk and flip chalkboard shown in Figure 10.1. This design was custom built for the nephew of a prolific maker, but now everyone can build it using these simple plans, with access to a hardware store and basic hand tools.

Figure 10.1 Wall-mounted work desk/chalkboard designed by CrazyCleverFollow. (CC BY-ND-SA) https://www.instructables.com/id/Wall-mounted-work-deskchalkboard/

Other examples include those that hack together commercial products. One of the most popular sites for this type of hacking is IKEAhackers (www.ikeahackers.net). This is a website with no affiliation with the commercial giant IKEA that can best be described as a fan-run website for taking IKEA's commercial offerings and making them better. For example, by combining a few IKEA offerings, you can create a transforming coffee table, as shown in Figure 10.2, known as the KNUFF. The KNUFF magazine file comes with two pieces, with one piece located inside the other, which you can use to change the shape of the table.

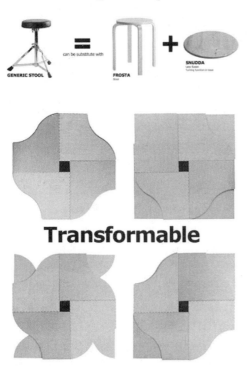

Figure 10.2 KNUFF transformable coffee table. Designed and shared, with permission, by J. Yap in Malaysia, 2012. IKEA Hackers, https://www.ikeahackers.net/2012/09/knuff-transformable-coffee-table.html

Good wood-based products do not need to come from new materials either. There is a thriving community of woodworkers creating unbelievably beautiful creations using waste pallets as their medium. Morning Chores (morningchores.com/pallet-projects) has cataloged more than 100 of them—everything from children's toy shelves and bookcases (like we have in my house) to beds, kitchen islands, desks, stools, sofas, outdoor spas, swings, and more. With a small amount of patience to deconstruct a pallet, you can usually get free wood and make unique furniture for your house that is truly your own. This is good for the environment as well as your wallet.

In addition, digital fabrication is starting to overlap with classic making activities like woodworking. For example, there are now a number of groups providing open-source plans for furniture that can be digitally fabricated (France, 2014) following distributed manufacturing (to be discussed in Chapter 12). In general, this computer numerical control (CNC) furniture design is cut from bulk thin materials using CNC routers and laser cutters, which allow for an expanding digitally manufactured open-source furniture movement. For example, SketchChair (www.sketchchair.cc) is a free open-source software tool that allows you to easily (1) design, (2) test, (3) export the cuts, and (4) build your own digitally fabricated furniture, as shown in Figure 10.3. After you make your own chair, you can upload your design to add it to a growing collection of open-source designs in the SketchChair Design Library. Then, anyone can download your chair design and make it, or if they do not have the CNC tools needed, they can send it to a company like Ponoko (www.ponoko.com), which is one of the first manufacturers that uses distributed on-demand manufacturing.

Other groups, such as OpenDesk (www.opendesk.cc), specifically use this business model for open-source furniture manufacturing. OpenDesk is an online marketplace that hosts independently designed furniture and connects its customers to local makers around the world. OpenDesk designs are largely Creative Commons, available to

Figure 10.3 The process of SketchChair is a free open-source software tool that allows anyone to design and build their own digitally fabricated furniture, along with an example chair. (CC BY-SA-NC) http://www.sketchchair.cc/

Figure 10.4 Lift standing desk designed and shared by Joni Steiner and Nick Ierodiaconou on Opendesk. (CC BY-SA-NC) https://www.opendesk.cc/lean/lift-standing-desk#make-it-yourself

download for noncommercial use (OpenDesk 2019). Designers can choose whether to make downloads available for free (which will increase distribution and uptake and is a nice way to provide value to individuals, charities, students, and schools without having an impact on commercial revenue) or charge for downloads. For example, the lift desk shown in Figure 10.4 is an adjustable-height workbench that goes from a seated level (738 millimeters) to bar level (912 millimeters) to a standing-desk height (1,095 millimeters). When you are working at the seated height, the frame provides two useful shelves. You can download the design files, cut out the pieces on your own CNC mill, or order a kit from someone else, including the makers who are part of the OpenDesk network. With this transforming desk system, you can avoid sitting all day and thus improve your health.

To get an idea of how the CNC furniture process works, see the design and assembly files for an open-source Silver Lining bed designed by AtFAB in Figure 10.5a; in addition, AtFAB produces the bed shown in Figure 10.5b fully assembled. This bed is constructed primarily from six 4- × 8-foot sheets of ¾-inch material and 36 fasteners. In addition, different sizes, including U.S. queen, U.S. king, and 1,800 ×1,800 millimeters, have been prototyped and tested.

This movement involves more than just small-scale wood projects. Open Source Wood (opensourcewood.com), sponsored by Metsä Wood, a major provider of wood products for construction, is actively recruiting architects and engineers to share their innovations in wood construction. The Open Source Wood initiative is using open innovation to facilitate knowledge sharing and growth in modular wood construction. This, in turn, will help drive

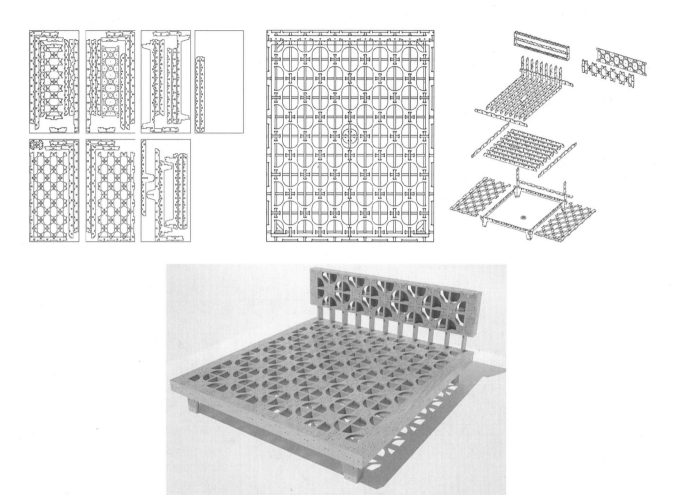

Figure 10.5   The Silver Lining Bed in the AtFAB line of customizable open-source furniture designed for distributed digital manufacturing. (CC BY-SA-NC) http://atfab.co/?portfolio=silver-lining-bed

sustainability and a circular economy because wood remains one of the most sustainable construction materials (Falk, 2009; Lehmann, 2012; De Araujo et al., 2016). At Open Source Wood, designers share their ideas so that others can use them to overcome the knowledge gap about modular wood design and building.

For example, Kate Rybenchuk's Sponge design is shown in Figure 10.6a. She is an architect from the Ukraine who designed a porous high-rise building as a combination of two almost-all-wood modular structures (Figure 10.6b). The details of the first module are shown in exploded view in Figure 10.6c. As can be seen in Figure 10.6b, the modules form a porous structure in the high-rise building with "holes"—

meant for terraces for greenery, as shown in Figure 10.6a.

Finally, you can put it all together. Whether you are living in a supercool open-source Sponge or a less exciting proprietary row house, you can get the measurements of your rooms (an easy download for the former, and less easy "tape measure and elbow grease" for the latter, or perhaps try some photogrammetry, as discussed in Chapter 3). Then, you can use the open-source SweetHome 3D software package (www.sweethome3d.com) to design the interior of your home by placing all your open-source furniture in the best locations. You can draw walls, rooms, windows, and furniture in SweetHome 3D in order to lay out

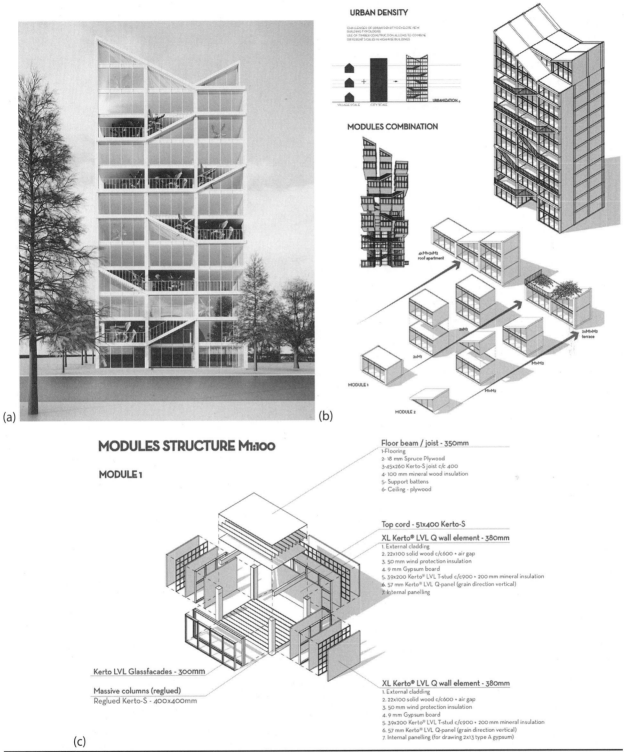

URBAN DENSITY

MODULES COMBINATION

(a) (b)

MODULES STRUCTURE M1:100

MODULE 1

Floor beam / joist - 350mm
1-Flooring
2- 18 mm Spruce Plywood
3-45x260 Kerto-S joist c/c 400
4- 100 mm mineral wood insulation
5- Support battens
6- Ceiling - plywood

Top cord - 51x400 Kerto-S

XL Kerto® LVL Q wall element - 380mm
1. External cladding
2. 22x100 solid wood c/c600 + air gap
3. 50 mm wind protection insulation
4. 9 mm Gypsum board
5. 39x200 Kerto® LVL T-stud c/c900 + 200 mm mineral insulation
6. 57 mm Kerto® LVL Q-panel (grain direction vertical)
7. Internal panelling

Kerto LVL Glassfacades - 300mm

Massive columns (reglued)
Reglued Kerto-S - 400x400mm

XL Kerto® LVL Q wall element - 380mm
1. External cladding
2. 22x100 solid wood c/c600 + air gap
3. 50 mm wind protection insulation
4. 9 mm Gypsum board
5. 39x200 Kerto® LVL T-stud c/c900 + 200 mm mineral insulation
6. 57 mm Kerto® LVL Q-panel (grain direction vertical)
7. Internal panelling (for drawing 2x13 type A gypsum)

(c)

Figure 10.6 (a) Rendering of Sponge, an open-source wood housing unit; (b) conceptual design of the two modules; (c) exploded-view details of the first module by K. Rybenchuk. (CC BY-4.0) https://opensourcewood.com/pages/Idea.aspx?i=130

a room. When you are done, you can create photorealistic images and videos with various light sources to view your design in three dimensions. This is a big bonus because you do not have to move the sofa back and forth across the living room in real life. In addition, this could save you a lot of time and money hiring an interior decorator or finding out too late that you purchased (or worse yet made!) a piece of furniture that cannot fit in your space.

References

De Araujo VA, Vasconcelos JS, Cortez-Barbosa J, et al. 2016. Wooden residential buildings: A sustainable approach. *Bulletin of the Transilvania University of Brasov. Forestry, Wood Industry, Agricultural Food Engineering II* 9(2):53.

Falk RH. 2009. Wood as a sustainable building material. *Forest Products Journal* 59(9):6–12.

France A. 2014. Open CNC Furniture | Make: DIY Projects and Ideas for Makers. https://makezine.com/2014/03/18/open-source-furniture-2/

Lehmann S. 2012. Sustainable construction for urban infill development using engineered massive wood panel systems. *Sustainability* 4(10):2707–2742.

Opendesk. 2019. Opendesk: Non-Commercial Use. https://www.opendesk.cc/about/non-commercial-use

Schwarz C. 2014. Open-Source Woodworking. Lost Art Press, https://blog.lostartpress.com/2014/09/04/open-source-woodworking/

CHAPTER 11

Making and Sharing Electronics

Arduino

As you will see in this chapter, there is a ton of open-source electronics projects to automate and control almost anything you like. One project in particular has made this possible: the Arduino. An Arduino (www.arduino.cc) is an open-source electronics platform based on easy-to-use microcontroller hardware and open-source software. Think of the Arduino as a lightweight brain enabling you to complete any automation-related task. Originally, it was developed as an electronics prototyping platform for design students in Italy by Massimo Banzi and David Cuartielles. They were frustrated that teaching design students the complicated engineering skills necessary to program a conventional microcontroller left little time to focus on design, their real passion. In addition, conventional commercial microcontrollers were obscenely expensive. The costs and complexities made automating things really challenging for all but well-funded professionals. Banzi and Cuartielles fixed both problems. Both the official Arduino and its open-source copies now cost a shockingly low amount. The Arduino Uno costs less than $10, and the Arduino Nano costs about $1. This makes even complicated electronics projects extremely cost-effective. The Arduino platform consists of a relatively simple board using Atmel's ATMEGA8 and ATMEGA168

microcontrollers and onboard input-output support. Thus, it is possible for you to drop the costs even further if you use the chips rather than full boards in your own electronics (see the next two sections).

However, for beginners, using the Arduino makes life easy. The Arduino team had the foresight to share its innovation using open-source principles that not only allowed all the rest of us to benefit but also ensured that they now have a worldwide collaborative team consistently helping make their designs even better. There is now a massive family of open-source microcontrollers based on the Arduino. They can all be programmed and used in the same way—you can choose the features you need to realize your project.

Arduinos have been used for hundreds of projects. We saw how they could be used to control drones in Chapter 8, and we will see how they can be used for 3D printers in Chapter 12. The power of the Arduino is in the ease with which, and the speed at which, you can use it to get an interactive project working at a very low price. For example, if you like the idea of making your own musical instruments, as we discussed in Chapter 5, but you want to make your instruments more "space age," consider fabricating an Arduino-controlled laser harp (Figure 11.1).

Figure 11.1 Arduino laser harp. Instructions by Yaroshka. (CC BY-NC-SA) https://www.instructables.com/id/Arduino-Laser-Harp-1/

If you really liked the cartoon movie *Wall-E* and want to build your own cute little robot, you can follow easy instructions developed by do-it-yourself (DIY) makers in Spain, as shown in Figure 11.2.

Or perhaps you've always wanted your own magic mirror to let you know who is the "fairest of them all." Luckily, an Arduino-powered magic mirror and photo booth have been developed for your building enjoyment (Figure 11.3).

Maybe you really want a Segway but can't afford the real thing or even a commercial off-brand version. You can make your own Arduino-powered Segway-like device using some old bike parts, motors, a battery, and, of course, an Arduino (Figure 11.4). The bottom line for Arduinos is that you can make anything that involves sensing and acting on the environment much easier than you could with electronic controls in the past.

Figure 11.2 DIY makers: *Wall-E*. (CC BY-NC-SA) http://diymakers.es/wall-e/

Figure 11.3   Alinke DIY magic mirror and photo booth, Arduino powered. (CC BY-NC) https://www.instructables.com/id/The-Magic-Mirror/

Figure 11.4   Stoppi71, Arduino segway. (CC BY-NC-SA) https://www.instructables.com/id/Arduino-Segway/

This is because the Arduino integrated development environment, or IDE for short, makes programming a microcontroller literally child's play. The IDE is software that consists of a standard programming language compiler and the boot loader that runs on the board. The software side of the Arduino builds on an earlier open-source software language (Wiring) and integrated development environment (Processing). Thus, the Arduino IDE language (syntax and libraries) is almost identical to C++ with some slight simplifications and modifications, and the Processing-based IDE. It is easy to adapt Arduinos to your projects because the Arduino is extremely flexible and its process is relatively easy to learn and use for beginners. Without any previous electronics experience, you should be able to get the basics in an hour (especially with help or a few YouTube videos) and get well into a project of your own in an afternoon.

The Arduino board is a lot like your brain. It is extremely powerful and useful, but for it to act on the environment, it needs peripherals.

No matter how great your brain is, without your sensors (i.e., eyes, ears, nose, etc.) and your actuators (i.e., hands and legs, etc.), your brain would not be that fun to play with. In the same way, the Arduino can sense the environment by receiving input from a long list of sensors (e.g., acceleration, chemical, pressure, electric current and potential, humidity, light, magnetic fields, temperature, vibration, etc.). Then, based on rules you set in the IDE, the Arduino can affect its surroundings by controlling a similarly long list of outputs (e.g., lights, heaters, motors, robot hands, and other actuators). The Arduino can act as a standalone minicomputer running robots or drones or be connected to your laptop via a USB port so that you can program it or collect information on the fly. Enormous libraries of prewritten and free software covering almost any kind of sensor or actuator that you could imagine or buy can be downloaded for free (arduino.cc/en/Reference/Libraries). Arduino software is truly cross-platform and runs not only on free and open-source GNU/Linux operating systems but also on Windows and Mac platforms.

To get started, I recommend Lady Ada's tutorials (www.ladyada.net/learn/arduino). She is the founder of Adafruit (an open-source electronics firm) and maintains an excellent tutorial library, which should be useful for getting anyone started. Besides the online tutorials, there are many good books about how to use Arduinos (Banzi, 2011; Margolis, 2011; Monk, 2012).

The Arduinos, by themselves with a few sensors and actuators, are pretty handy, but they start to gain real powers with plug-and-play expansions called *shields*. Shields are an extremely important feature made possible by the standard way that connectors are exposed on the larger Arduino boards (Uno and bigger), allowing the board to be connected to other boards for specific tasks. Some shields communicate with the Arduino board directly over various pins, but many shields are individually addressable via an inter-integrated circuit (I²C) serial bus, allowing many shields to be stacked and used in parallel.

For a list of the more than 300 Arduino shields available and over 500 more in the queue as of this writing, see the Arduino shield list (shieldlist.org) maintained by Jonathan Oxer. For example, the SparkFun Musical Instrument Shield is an easy way to add Musical Instrument Digital Interface (MIDI) sound to your next Arduino project. You simply connect a speaker, stereo, or pair of headphones to the $\frac{1}{8}$-inch stereo jack on the shield and pass the proper serial commands to the integrated circuit (IC), and you'll be enjoying music. The shield contains two large tone banks including various piano, woodwind, brass, synthesizer, sound effects, and percussion sounds. The shield is also capable of playing several tones simultaneously (up to a maximum polyphony of 31 sounds). You can download the example code to add music to any

project and work on your composition skills in addition to your electronics skills. Chapter 13 will discuss other shields used by scientists.

Finding and Sharing Open-Source Electronics

Arduino is just one single family of microcontrollers—there are many, many more! Open-source electronics is one of the most mature subfields of open hardware. You can find almost any circuit you want in open-source circles. There are several repositories that you can use to download open-source electronics, such as the Open Circuits Institute (opencircuitinstitute.org), Open Electronics (www.open-electronics.org), Open Hardware Repository (ohwr.org), and Open Circuits Wiki (www.opencircuits.com/Projects), as well as Hackster (www.hackster.io), Codemade (www.codemade.io), and Hackaday (hackaday. com). All these websites allow both users and developers to discover and share open-source projects. There are also commercial offerings. Kitspace (kitspace.org) is a place to share ready-to-order electronics designs. In addition, there are several open-source electronics companies, including Sparkfun (www.sparkfun.com), Adafruit (www.adafruit.com), Tiny Circuits (tinycircuits.com), and Seeed Studio (www.seeedstudio.com).

Finally, you do not even need to rely on new electronic components to populate these open-source designs. Although the Arduino is largely plug and play, to build your own circuits, you will need to solder (www.instructables.com/id/How-to-solder). Learning to solder through-hole components and butt connections is an essential skill. You use a soldering iron that heats up solder and then allows you to melt and solidify the solder to make a permanent electrical connection between your components. This skill can be learned in a few minutes, but you will get better with more practice. After you know your way around a soldering iron, you can also disassemble broken electronic devices for the parts and then use them to make new creations. This is not only good for the environment but also can save you a lot of money.

Open-Source Software for Electronics Design

If you really get into open-source electronics, you may wish to make your own completely new electronics boards or build new capacities into other people's designs. To do this, the "gold standard" in open-source software is the suite for electronic design automation (EDA) called *KiCad* (www.kicad-pcb.org). KiCad handles schematic capture with a schematic editor that lets you create a design without any limits or paywalls that force you to pay money to unlock features. KiCad provides an official library for schematic symbols, and the built-in schematic symbol editor helps you to rapidly make and add your own designs.

You can make simple one-layer printed circuit boards (PCBs) with ease, but KiCad also makes professional PCB layouts with up to 32 copper layers. It also has a push and shove router that is capable of routing differential pairs and interactively tuning trace lengths. KiCad provides Gerber output, which is an open ASCII vector format for PCB designs. Gerber is the de facto standard used by PCB industry software to describe the PCB images: copper layers, solder mask, legend, drill data, and so on. You may need Gerber output if you decide to have a professional make your board after you have finished the design (explained in the next section).

Finally, KiCad includes a 3D viewer so you can inspect your design in an interactive canvas. This is important because, in a 2D view, you cannot rotate and pan around to inspect details (or hide and show features) that may ruin the functionality of your board. KiCad runs on Windows, Linux, and Mac operating systems and is licensed under GNU GPLv3, which means you are free to use it in any way you like as long as you share back to the community.

Another useful piece of open-source software for helping share your electronics projects with others is Fritzing (fritzing.org). Fritzing is an open-source hardware initiative that makes electronics easily accessible for almost anyone. I often use it to provide wiring diagrams to show people how to make a physical prototype if they have the board and the components. Besides the software, Fritzing has a robust community website and commercial services in the spirit of Processing and Arduino, which fosters a creative ecosystem. In the Fritzing community, you can document your prototypes, share them with others, get materials to teach electronics in your classroom, and even lay out and manufacture professional PCBs.

Ordering Boards

There are many businesses that provide PCB designers with services to make their electronics projects a reality. If you are just starting out and want either a prototype or a few PCBs, then you should look at the smaller manufacturers that cater to hobbyists and the open-hardware community. In the United States, one of the popular services in the open-hardware community is OSH Park (oshpark. com) for prototyping, hobby design, and light production. The company uses purple solder mask over bare copper (SMOBC) and an electroless nickel immersion gold (ENIG) finish, which works for lead-free reflow processes and is Restriction of Hazardous Substances Directive (RoHS) compliant. What this means is that OSH Park produces high-quality bare PCPs in purple! The boards are made in the United States and ship free anywhere in less than 12 days. OSH Park's costs are pretty straightforward, based on the number of layers and the number of square inches you need.

You may also want to consider working directly with an open-hardware company such as Seeed Studio (www.seeedstudio.com/fusion .html), which is a hardware innovation platform tailored for makers. The company offers the Seeed Fusion Service, which provides one-stop prototyping services for PCB manufacture and assembly, as well as other customized electronic and mechanical services such as CNC milling, 3D printing, and PCB layout services. The company provides a handy online price calculator, where you can make your selections for the type of board you want and get an immediate quote with a 100 percent quality guarantee.

If you have done your design in Fritzing and you need the PCB, parts for it, or stencils, Fritzing Fab, run by AISLER (aisler.net/ partners/fritzing), is an easy option. This is a full-service shop that can get your PCB to you in seven business days worldwide. Other options are outfits such as JLCPCB (jlcpcb.com), which is the JiaLiChuang Company (Hong Kong) and the largest PCB prototype enterprise in China. JLCPCB specializes in quick PCB prototyping and small-batch PCB production. The company has decent quality and extremely low prices (e.g., only $2 for a one- to two-layer 100- × 100-millimeter PCB). Similarly, if you want guaranteed 100 percent error-free boards, Advanced Circuits (www.4pcb.com) is one of the largest outfits in the world.

PCB Making

You may want to do everything yourself, or perhaps you have been making so many PCB designs that even the best deals from the PCB fabricators in the preceding section are burning

a hole in your wallet. Luckily, the open-source community provides several options for making your own PCBs.

Old School

The easiest way to make your own board is to breadboard it or use a conductive-ink pen to draw traces on a simple circuit. You can climb the ladder of sophistication a little higher by directly drawing the whole PCB layout on copper board using a black permanent marker (e.g., a Sharpie). First, draw your circuit using pencil, and when you are happy with it, trace it with a marker. Then, etch the unmarked copper away with a copper etching chemical such as ferric chloride or hydrogen peroxide. For more complex circuits, it is better if you use design software (as discussed earlier), print out the layout on glossy photo paper, and transfer it to the copper board using an iron. Then, the same etching process and possibly drilling through-holes will enable you to make a good PCB. Instructables (www.instructables.com/id/How -To-Make-A-PCB-PCB-Making-Guide or www.instructables.com/id/How-to-make-PCB -at-Home) has several useful tutorials that can walk you through the process.

Using a 3DP+ Mill

The open-source 3D printers (3DPs) that we discuss in Chapter 12 are extremely versatile and can be converted into moving-stage printers or substrate-moving robots (Zhang et al., 2016; Pearce, 2017). This feature is useful for electronics manufacturing. After converting a Delta RepRap into a stage printer, a copper-coated PCB blank can be attached to it. Then, a fixed tool is secured in place by a structure made of rectangular extruded aluminum that replaces one of the vertical boards on the delta robot. Tool heads such as a mill spindle can be secured to the mount by a magnetic mount

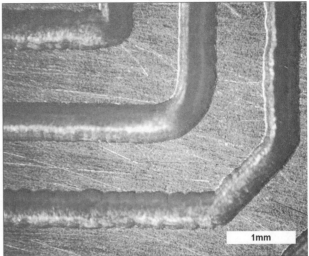

Figure 11.5 Digital micrographs of traces milled in PCBs produced by a commercial PCB mill (left) and the convertible 3D printing platform (right). (CC BY-SA) https://www.appropedia.org/File:Os-pcb.png

consisting of three rare earth magnets. A mill spindle holder with a 200- to 400-W direct current (DC) spindle motor that turns at 12,000 rpm when supplied with 48 V of power can be used to print out reasonable PCBs (Anzalone et al., 2015). PCBs designed in KiCAD (as discussed earlier) can be converted to G-code using the open-source pcb2gcode (sourceforge.net/projects/pcb2gcode/). Figure 11.5 shows digital micrographs of traces milled in PCBs produced by a commercial PCB mill (left) and the convertible 3D printing platform (right). Clearly, the 3D printer operating as a mill shows more chattering than a CNC machine meant only for milling, but it is still more than adequate for the vast majority of hobby-level PCB milling projects.

Dedicated Open-Source Mill

If you are interested in making even more sophisticated PCBs (or if you are going to do a lot of them), you can build an open-source mill meant specifically for PCBs. There are several open-source PCB mill designs on Open Builds (openbuilds.com) and in open-

hardware/electronics repositories. An Open Source Ecology–designed distributed 3D (D3D) robotics system can also be converted into a dedicated PCB mill (Oberloier and Pearce, 2018), as shown in Figure 11.6. You can compensate for any motion inaccuracies of the open-source mill with the Open Circuit Institute's open-source Copper Carve, which enable the machine to

Figure 11.6 Belt-driven open-source circuit mill developed around low-cost 3D-printer components (Oberloier and Pearce, 2018). (GNU FDL) https://www.appropedia.org/File:OScircuitmill.png

achieve a motion resolution of 10 microns (less than the thickness of a human hair). This should be more than adequate for the vast majority of circuit designs. The open-source mill is at least five times less expensive than all commercial alternatives. Thus, if you plan to make more than 20 boards, the open-source mill pays for its material costs.

Open-Source Security

One of the counterintuitive benefits of embracing open source is that it can make you, your family, and your company more secure. It is perhaps easiest to see this with the hysteria surrounding the recent rash of hardware hacks. Hardware hacks are particularly scary because they trump any software security safeguards. For example, they can render all accounts on a server password-less. Fortunately, we can benefit from what the software industry has learned from decades of seemingly tireless and prolific software hackers: Using open-source techniques can make a system more secure (Hoepman and Jacobs, 2007). How can this be?

Imagine that you are a James Bond–like 007 agent holding classified documents. Would you feel more secure locking them in a safe whose manufacturer keeps the workings of the locks secret or locking them in a safe whose design is published openly so that everyone (including thieves) can judge its quality—thus enabling you to rely exclusively on technical complexity for protection?

The first approach might be perfectly secure— you simply don't know. But why would you trust any manufacturer that could be compromised now or in the future? If the information is really important, it is insane to trust it to the "black box." In contrast, the open system is almost certain to be secure, especially if enough time has passed for it to be tested by multiple companies, governments, and individuals. Thus, counterintuitively, security is one of the core

benefits of open-source technology development (Lynch, 2015).

While open source is not inherently more secure, it allows you to verify security yourself (or pay someone who is more qualified to do so for you). With closed-source programs, you must trust, without verification, that a program works properly. To quote former President Reagan: "Trust—but verify." This is a far better way to approach security, whether online or not. The bottom line is that open source allows you, the user, to make more informed choices about the security of a system. With closed-source software, you simply have trust alone.

This concept of "trust but verify" also holds true for electronic devices. Most electronics customers have no idea what is in their products, and even technically sophisticated companies like Amazon may not know exactly what is in the hardware that runs their servers because they use proprietary products that are made by other companies. This can be really scary. Recently, in an incident reported by Robertson and Riley in *Bloomberg* (2018), Chinese spies used a tiny microchip, not much bigger than a grain of rice, to infiltrate hardware made by SuperMicro (the Microsoft of the hardware world). These chips enabled outside infiltrators to access the core server functions of some of America's leading companies and government operations, including Department of Defense data centers, CIA drone operations, and the onboard networks of Navy warships. Operatives from the People's Liberation Army or similar groups could have reverse-engineered or made identical or disguised modules (in this case, the chips looked like signal-conditioning couplers, a common motherboard component, rather than the spy devices they were).

Having the source available helps customers much more than hackers, because most customers do not have the resources to reverse-engineer the electronics they buy. Without the device's source, or design, it's difficult to determine whether or not hardware has been hacked. This is where open hardware and distributed manufacturing (where you can actually inspect the core source and make sure you are getting it) add to security.

Open-source hardware and distributed manufacturing could have prevented the Chinese hack that rightfully terrified the security world. Organizations that require tight security, such as military groups, could then check the product's code and bring production in-house, if necessary. In the end, no one really wants to be spied on. With open hardware, all users can increase the security of their devices. Not long ago, you had to be an expert to make even a simple breadboard design. Now, as shown earlier, with open-source mills for boards and electronics repositories, small companies and even individuals can make reasonably sophisticated electronic devices. While most builders are still using black-box chips on their devices, this is also changing as open-source chips gain traction (Waite, 2010). Creating electronics that are open source all the way down to the chip is certainly possible—and the more besieged we are by hardware hacks, perhaps it is even inevitable. Companies, governments, and other organizations that care about cyber security should strongly consider moving toward open source—perhaps first by establishing purchasing policies for software and hardware that make the code accessible so that they can test for security weaknesses. If you are worried about hackers or identify theft even a little bit, open hardware provides some support, and in general, making your own electronics costs significantly less than buying black-box electronics. Although challenges certainly remain for the security of open-source products (Wen, 2017), the open-hardware model can help enhance cyber security—from the Pentagon to your living room.

References

Anzalone G, Wijnen B, Pearce JM. 2015. Multi-material additive and subtractive prosumer digital fabrication with a free and open-source convertible delta RepRap 3-D printer. *Rapid Prototyping Journal* 21(5):506 –519, doi: http://dx.doi.org/10.1108/RPJ-09-2014-0113

Banzi M. 2011. *Getting Started with Arduino*, 2nd ed. Make: Santa Rosa, CA.

Hall C. 2018. Survey shows Linux the top operating system for internet of things devices. *IT Pro*. https://www.itprotoday.com/iot/survey-shows-linux-top-operating-system-internet-things-devices

Hoepman JH, Jacobs B. 2007. Increased security through open source. *Communications of the ACM* 50, no. 1: 79-83.

Kerner SM. 2018. Linux Foundation on track for best year ever as open source dominates. *ServerWatch*. https://www.serverwatch.com/server-news/linux-foundation-on-track-for-best-year-ever-as-open-source-dominates.html

Lynch J. 2015. Why is open source software more secure? *InfoWorld*. https://www.infoworld.com/article/2985242/why-is-open-source-software-more-secure.html

Margolis M. 2011. *Arduino Cookbook*. O'Reilly Media, Sebastopol, CA.

Monk S. 2012. *Programming Arduino: Getting Started with Sketches.* McGraw-Hill, New York.

Oberloier S, Pearce JM. 2018. Belt-driven open source circuit mill using low-cost 3-D printer components. *Inventions* 3(3):64. https://doi.org/10.3390/inventions3030064

Pearce JM. 2017. Impacts of open source hardware in science and engineering. *National Academy of Engineering: The Bridge* 47(3): 24–31.

Robertson J, Riley M. 2018. China used a tiny chip in a hack that infiltrated U.S. companies. *Bloomberg*. https://www.bloomberg.com/news/

features/2018-10-04/the-big-hack-how-china
-used-a-tiny-chip-to-infiltrate-america-s-top
-companies

Vaughan-Nichols SJ. 2018. Supercomputers:
All Linux, all the time. ZDNet. https://www
.zdnet.com/article/supercomputers-all-linux
-all-the-time/

Waite E. 2010. Using open source designs to
create more specialized chips. *Wired*. https://
www.wired.com/story/using-open-source
-designs-to-create-more-specialized-chips/

Wen S. 2017. Software security in open source
development: A systematic literature review.
In 21st Conference of Open Innovations
Association (FRUCT). IEEE, Conference
Location: Helsinki, Finland; pp. 364–373.
https://doi.org/10.23919/FRUCT.2017.825020

Zhang C, Wijnen B, Pearce JM. 2016. Open-source
3-D platform for low-cost scientific instrument
ecosystem. *Journal of Laboratory Automation*
21(4):517–525.

CHAPTER 12

Making and Sharing Digital At-Home Manufacturing

Now we will look at the three big power tools of digital at-home manufacturing: the 3D printer, the laser cutter, and the CNC mill. With each of these devices, if you can make one or get access to one, you suddenly have the ability to make millions of products for yourself. The designs for these products can come from databases of millions of open-source ideas. Perhaps more excitingly, the designs can be ones that you have dreamed up and would like to realize in real life. For each of these technologies, there are accessible starting points in the open-source ecosystem, yet they also have evolved enough that you can fabricate very sophisticated tools that can match product fabrication in industry. In the end, each of these tools is really a different type of end effector mounted on a 2D or 3D platform computer-controlled robot. The end effectors do either additive manufacturing (where you add material as you fabricate something) or subtractive manufacturing (where you take material away from a solid block). Once you understand one, you can jump back and forth between additive and subtractive manufacturing to make just about anything!

RepRap 3D Printer

3D printing and additive manufacturing (AM) are actually old technologies. You would never know that, though, because they only gained popular attention when the open sourcing of the technology dropped the costs so that they became accessible to everyone. The recent development of open-source 3D printers makes the scaling of mass-distributed AM of high-value objects technically feasible for everyone from open-source plans (Gershenfeld, 2005; Bradshaw et al., 2010; Pearce et al., 2010; Weinberg, 2010; Cano, 2011). These RepRaps (self-REPlicating RAPid prototypers [www.reprap.org]) were developed in 2005 by Dr. Adrian Bowyer, a professor of mechanical engineering at the University of Bath. Bowyer has a particularly clear view of the implications of what a general-purpose self-replicating manufacturing machine means for humanity and how the project was able to literally evolve under the open-source paradigm (Bowyer, 2004; Sells et al., 2007; Arnot, 2008; Jones et al., 2011; Hodgson, 2013). This is why the major early variants of the RepRap are all named after biologists: Darwin, Mendel, Huxley, and Wallace. The RepRap evolutionary family tree has gotten completely out of control, with hundreds of variants of RepRap 3D printers evolving all over the world (for some examples, see Figure 12.1).

Figure 12.1 Screen capture of a tiny fraction of the major evolutionary paths for the RepRap 3D printer. (GNU FDL-1.3) https://reprap.org/wiki/RepRap_Machines

The RepRap project has created 3D-printing machines that can manufacture approximately half of their own parts from sequential fused deposition of a range of polymers and common hardware. Some versions can do even more of their own parts. They can also make millions of products for you—all developed and licensed for free by others. Likewise, they can bring ideas to life that you have just thought of and that you can share with the global community. The RepRap is made up of a combination of printed mechanical components, stepper motors for 3D motion and extrusion, and a hot end for melting and depositing sequential layers of polymers, all of which are controlled by an open-source microcontroller such as the Arduino (discussed in Chapter 11). The extruder takes in a filament of the working, normally thermoplastic material, melts it using electric heating, and extrudes it through a nozzle. The printer uses a three-coordinate system, where each axis involves a stepper motor that makes the axis move until it reaches a limit switch for homing. The printing process is a sequential layer deposition where the extruder nozzle deposits a 2D layer of the working material, then the Z (vertical) axis will raise, and the extruder will deposit another layer on top of the first, building a 3D object in 2D layers, one at a time.

Because the open-source ethic is so deeply rooted in the 3D-printing community, many of the consumer products you normally buy are already among the millions of free predesigned products. You can download the designs and save a lot of money. My group has shown in studies on a self-built 3D printer (Wittbrodt et al., 2013) and an out-of-the box 3D printer (Petersen and Pearce, 2017) that peer-to-peer sharing "prosumers" (producing consumers) gain an incredible return on investment—more than 100 percent at a minimum and more likely approximately 1,000 percent—by 3D printing products to offset purchases, only using the printer once a week.

Your "investment" in this case is an open-source 3D printer derived from the RepRap. An open-source 3D printer costs less than $200 if you build it yourself, or $1,250 to $2,500 if you buy one preassembled. You can get a decent open-source RepRap partially assembled for $200 to $400 as of this writing.

Downloading a predesigned open-source file and clicking "Print" is pretty easy. Here, I will lay out the steps you need to take to make a product no one has open-sourced yet, which is a bit trickier. In addition, *I will show you how to make it for less than the sales tax on an equivalent commercial proprietary product.*

My daughter recently started Nordic skate skiing, which means I suddenly needed a ski waxing rack. To get an idea of what this means, see Steve Foreman's excellent how-to video (www.youtube.com/watch?v=YX-wiE3oItg) on ski waxing. We went to the local outdoor sports store, and the cost of a rack was a whopping $125. It was so expensive that the store had the box locked in a glass display case. All I needed to do, however, was see the basic physical makeup of the product to make a better one using the open-source design method. To demonstrate the method, I will walk you through how I made my daughter a fan art–based Wonder Woman–themed ski waxing rack for less than $5.

Here are five steps you can use to leverage digital manufacturing open-source hardware such as the RepRap to truly obliterate the costs of consumer goods:

1. **See if anyone has designed the product you want already.** I recommend using Yeggi (yeggi.com), which is a 3D-printable search engine to find premade free designs. I knew that at least one ski waxing rack was available because one of my Michigan Tech students had designed an excellent set of ski waxing vises on Libre3D (libre3d.com). Jerry Trantow and someone named Mosquito4 had made Nordic ski waxing table fittings and a complete ski vise for downhill skiing and posted them on Thingiverse (www.thingiverse.com). I didn't want a whole table or a vise, but I wanted a system for Nordic cross-country skis. I just wanted to use the kitchen table, so I needed

to make something new. Even though these designs were not exactly what I wanted, they provided me with some good ideas on how I could do my design. To make it interesting, I chose to design maximizing the printable components to minimize the cost. This is following the RepRap tradition—minimize all parts that are not 3D printable.

2. **Stand on the shoulders of giants and use similar designs for components of what you want.** If no one has yet made what you want, then see if they have made any of the parts. In my case, I needed some C- or G-clamps to hold the skis to the kitchen table. It turns out that Johann Joe had made and shared some pretty nice fully printable ones (www.thingiverse.com/thing:1673030). Unfortunately, Johann only shared the STL files. These are the files you need to print with and represent the outer surface of a 3D object. Ideally, Johann would have shared the source code for the design, using an open-source computer-aided design (CAD) package to make the design. But no matter, most CAD packages allow you to edit the STL file as well.

3. **Use an open-source CAD package such as OpenSCAD (www.openscad.org) to adapt the design for your needs.** I can't say enough good things about OpenSCAD and the good people who have contributed to the code base and free tutorials. It is a powerful parametric script-based CAD package. What this means is that it allows you to computer code 3D objects. If you know a little bit of coding, you will be right at home with OpenSCAD. I took some measurements of my daughter's skis to ensure customized designs. All measurements in OpenSCAD are in millimeters. I used the Import command

to bring the core part of Johann's G-clamp into OpenSCAD, and then, because I wanted the result to look nice, I also used a rounded-edge cube library (www.thingiverse.com/thing:470007) created by P. Towalski to build up the parts I wanted.

I made three clamps, the middle one that holds the ski by the standard binding and two tilting holders. The middle holder is the only one with any metal. I found the metal wire I needed from a broken clothes hanger from my daughter's closet. I was careful to make all the parts easy to print from the surface of the print bed while minimizing overhangs. This meant that the middle piece needed to be in two parts. I did this because I wanted to minimize the use of plastic (you can do overhangs even with a single-nozzle 3D printer, but then you waste plastic building up support and have to spend time cleaning up your parts afterward). Here, I had the ugly overhang hidden in the square hole of the middle piece (Figure 12.2).

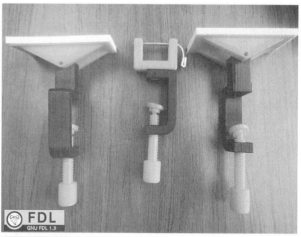

Figure 12.2   The free and open-source ski waxing rack system. (GNU FDL-1.3)

4. **Print on an open-source 3D printer.** I printed all the parts in polylactic acid (PLA) on an open-source Lulzbot commercial RepRap 3D printer (Figure 12.3). PLA is the most common commercial 3D-printing plastic. It is a biopolymer that is both recyclable and compostable. Buying a prebuilt open-source 3D printer such as a Lulzbot or Prusa is easier for most people, but all the steps would be the same on a RepRap you built yourself from scratch or on a low-cost kit such as the open-source Creality Ender 3.

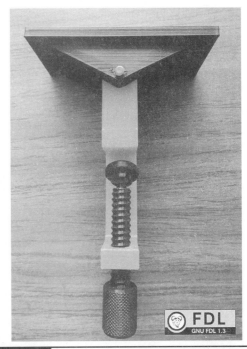

Figure 12.4　Rivet close-up. (GNU FDL-1.3)

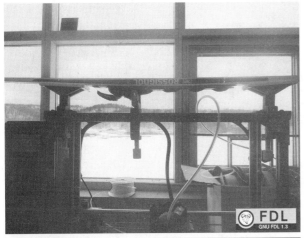

Figure 12.3　The Lulzbot Taz, which was used as a test bed to print the ski waxing rack. (GNU FDL-1.3)

To put the pieces together, I used filament as rivets (Figure 12.4), following a great tutorial (www.youtube.com/watch?v=WqNxYHWpYko) by Jason Welsh, who needed them for his line of open-source 3D-printable action figures. You can use a soldering iron—but if you don't have one, a clothes iron set to medium works just fine.

I didn't want the top of the skis to get scratched when in the mount, so I printed some covers with NinjaFlex filament. Later, I found that this was not at all necessary. Finally, to add a super-heroine flourish, I borrowed the Wonder Woman design (Figure 12.5) from a design of a cookie cutter made by g33kgirl.

5. **Share!** All the ski waxing system files can be downloaded for free from MyMiniFactory (www.myminifactory.com/object/nordic -skate-skiing-waxing-rack-28893).

I was telling a friend on the ski trails about my great success (the economics are in the next section), and she asked if she could have a Thor one for her family (all boys). With the power of the global sharing community, she could do it with about 5 minutes of additional design time. I looked up an OpenSCAD hammer for Thor. As luck would have it, the legendary Vik Oliver (the guy who built the first RepRap child)

Figure 12.5   Wonder Woman NinjaFlex cover. (GNU FDL-1.3)

had SCADed up a pretty nice Thor hammer (www.thingiverse.com/thing:7985) to be used as a pendent. Vik had modeled his version on a design posted on Wikipedia from an artifact in the Swedish Museum of National Antiquities, which someone else had uploaded there in the Creative Commons. Are you starting to understand how awesome the sharing world is? Because my second version should always be better than the first, this time I cut the hammer directly out of the two sides, after mirroring it twice in OpenSCAD, and thus had the design without the added costs and relatively slow printing time of NinjaFlex (Figure 12.6).

Maker Economics

Both the Wonder Woman and Thor hammer versions of the ski waxing system worked great, look cooler, and destroyed the cost of a generic commercial mass-manufactured product. By how much?

Commercial PLA filament costs about $0.025/g, and NinjaFlex costs $0.086/g, which means that the Wonder Woman hard parts cost $3.25 and the soft pads cost $1.73. All in all, my completely 3D-printable ski waxing system printed for $4.98 for the Wonder Woman version. This a whopping 96 percent savings from the commercial system. *In fact, because the sales tax in Michigan is 6 percent, my design is not only cooler, but the open-source 3D-printable ski waxing system also costs less than the cost of the taxes on the commercially available system.* The Thor hammer version is even less expensive because it is made only of PLA and costs $3.33. This is not even as low as I could go because I could have used recycled plastic from a recyclebot (www.appropedia.org/Recyclebot) to push the cost to a few pennies, as you will see in Chapter 15. Equally interesting is the fact that by printing out two sets of ski waxing racks (which could be easily done in a weekend day), I had already justified the cost of an entry-level 3D printer.

Figure 12.6   Thor's hammer version of the ski waxing system. (GNU FDL-1.3)

The Bottom Line

If you do not live in a tundra (or in my case, a taiga [snow forest]), you may not have the need for a ski waxing rack, but that is part of the point. Whatever consumer products you need, chances are that most of them already have an open-source 3D-printable version available. For uncommon products such as a Thor-themed ski waxing rack that no one has thought of yet, you can use this method. If you share it, it makes it more likely that others will join the open-source community when they make their less-common creations as well. We all win!

The bottom line is that with a little practice working in OpenSCAD or another open-source CAD package (e.g., FreeCAD [www.freecadweb .org]) and the foundation of designs already available in the free-sharing 3D-printing community, you can save on specialty goods even if no one has gotten around to designing them yet. The open-source approach to distributed manufacturing of consumer goods is real. It saves money—lots of money. It is here now, and you can be a part of it. Although open-source 3D printing has really exploded in popularity, you can use the same open-hardware techniques for subtractive manufacturing as well with both laser cutters and mills.

Open-Source Laser Cutting and Engraving

The easiest way to get started with laser cutting and engraving is to simply swap out the head on your RepRap with a laser. Several of my students have modified the 3D printers they built in my class to do this for the laser cutter or 1-watt laser engraver shown in Figure 12.7. Other RepRappers have moved into lasers, using axRap's design of an AxCut laser cutter (axrap. blogspot.com), and Jake's Workshop has also published a method for turning any 3D printer into a 2.5-watt laser engraver, as shown in Figure 12.8. OpenBuilds (openbuilds.com/threads/

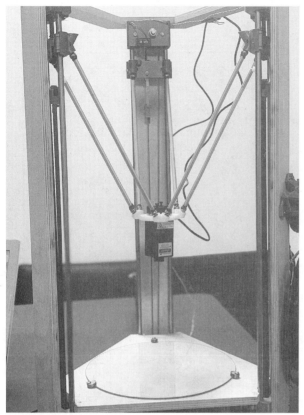

Figure 12.7   A modification of a delta-style RepRap (Athena II) to convert it to a 1-watt laser engraver. (GNU FDL) https://www.appropedia.org/Athena_II_Laser_Engraver_mod

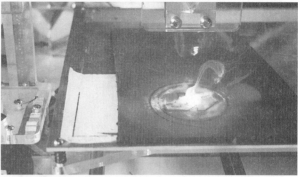

Figure 12.8   Laser engraver/cutter in action on a converted 3D printer. (CC BY-NC-SA) https://www.instructables.com/id/Convert-Any-3D-Printer-to-Laser-Engraver/

and an optical-beam delivery subsystem. It has a maximum feed/seek rate of 6,000 millimeters per minute, which makes for fast cutting. It can do all your favorite types of cutting and engraving.

Figure 12.9   The open-source Lasersaur, whose reference design is CC BY-SA, software is GPLv3, and manual is CC BY-SA. https://github.com/nortd/lasersaur/wiki

openbuilds-table-top-50w-co2-laser-cutter-engraver.14364) offers a dedicated 50-watt carbon dioxide (CO_2) laser cutter and engraver.

Another approach is to get serious with a large-format laser cutter. There are several options. The mother of open-source laser cutters is the 120-watt Lasersaur (github.com/nortd/lasersaur/wiki), which stands at $1,700 \times 1,170 \times 360$ millimeters and gives you a work area of 2×4 feet with a resolution of 0.1 to 0.03 millimeters (240–840 dots per inch [dpi]; Figure 12.9). Laser cutters focus the energy of a laser beam onto a small spot to control the energy density at a given time. The Lasersaur control software does this by running a combination of robotic actuators and sensors, a laser resonator,

Another approach is shown in Figure 12.10. A fablab built a LaserDUO (laserduo.com), which integrates two different laser sources (a 130-watt CO_2 laser and a 75-watt yttrium-aluminum-garnet (YAG) laser) over its $1,500-\times 1,000$-millimeter working area (with 500 millimeters in Z-axis) and thus greatly expands its ability to both laser cut and engrave a wide range of materials on large parts with just one

Figure 12.10   LaserDUO, an open-source dual laser cutter/engraver (inset: products of the fablab). (CC BY-SA-NC)
http://www.laserduo.com/

machine. The laser can do wood, plexiglass, polyoxymethylene (POM)/Derlin, medium-density fiberboard (MDF), cardboard, and fabrics, including cloth and leather. With the YAG laser, however, you can also cut steels, brass, copper, aluminum, magnesium, and even marble. (You can also transform your laser cutter into a 3D printer with 1 cubic meter of printing volume.)

Open-Source Mills

Computer numerical control (CNC) devices are used to fabricate physical objects with a high degree of precision, and if they are mills, they do it with subtractive manufacturing—that is, cutting away what you do not want in your object from a block of material. I talked about

the open-source CNC circuit mill and an upside-down delta 3D printer with a mill attachment for making electronic printed circuit boards in Chapter 11. There is a lot more you can do with a CNC mill, however. Most CNC mills feature a gantry-mounted cutting tool (such as a router) that can move in two or more directions. The open-source community has provided a wide range of such tools. For example, if you have access to a laser cutter (or some hand tools and patience), you can follow instructions to make a $700 three-axis CNC mill called *DIYLILCNC* (diylilcnc.org), as shown in Figure 12.11. If you do not have access to a laser cutter, you can purchase laser-cut panel kits from the company or give the design files to any online laser-cutting service.

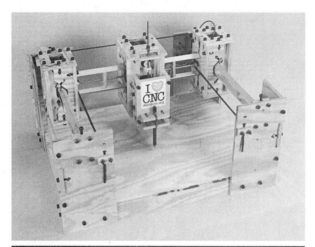

Figure 12.11 The big shoulders of DIYLILCNC. (CC BY-SA) diylilcnc.org

There are many other options for CNC milling. Another example is the OpenBuilds Mini Mill, which is a desktop-sized CNC milling plate, parts maker, and 3D carving machine, as shown in Figure 12.12. The OpenBuilds Mini Mill can be assembled in 2 to 4 hours depending on your skills. It has an accuracy of 0.05 to 0.1 millimeter, a workable material height of 60 millimeters, and a 120- × 180-millimeter work area. Another approach to open-source milling is to make a 3D stage using 3D-printed parts. Some of the simplest designs involve mounting an existing tool such as a Dremel on a 3D stage. The do-it-yourself (DIY) Dremel CNC (www.thingiverse.com/thing:3004773) runs on an Arduino (see Chapter 11) using primarily 3D-printed parts and, of course, a Dremel. If you want to go bigger, there are several machines designed by Ryan Zellars that might be what you are looking for. First is the Mostly Printed CNC (MPCNC) BURLY C-23.5 mm outside diameter (OD) shown in Figure 12.13. The only special tools you need are a hacksaw and a drill, along with any open-source fused filament fabrication (FFF) 3D printer. It can be easily expanded to almost any length, width, and

Figure 12.12 OpenBuild's Mini Mill and printed objects. (Public domain) https://openbuilds.com/builds/openbuilds-minimill.5087/

Figure 12.13 Mostly Printed CNC (MPCNC) BURLY C-23.5 mm OD. (CC BY-SA-NC) https://www.thingiverse.com/thing:724999

depth depending on the rigidity required. The shorter the lengths and the better the materials, the more rigid and accurate will be your CNC milling. Figure 12.13 shows an MPCNC built using a hardware store electrical metal tubing (EMT) conduit (in the United States, it is sold as ¾-inch inside diameter [ID]). You can upgrade to stainless steel tubing. The MPCNC can mill most woods and plastics, as well as aluminum and even steel (although for metals, you will want to work over a smaller area). As with any 3D motion platform, you can switch the spindle out with a drag knife, a laser, or a 3D printer extruder to make a prototyping multitool. Finally, if you want to use whole sheets, you can use Zellar's horizontal design (LowRider2 CNC [www.thingiverse.com/thing:3064287]) or the Maslow, which is a $500 large vertical-hanging (4- × 8-foot) CNC cutting machine designed to let you cut big, useful things out of wood and other flat materials (Figure 12.14). The Maslow team has created Maslow Create, an open-source community cloud CAD program for cooperative design (github.com/MaslowCNC/Maslow -Create) and the Maslow Community Garden (maslowcommunitygarden.org). The latter is a sharing website for CNC designs that sports everything from furniture that you might expect (e.g., chairs, coat hangers, beds, and picnic tables) to truly creative wonders, from a Santa sleigh to a full tiny house! With mills such as this, you can scale any 2D image you see on the internet to 4 × 8 feet with Inkscape and make everything in your house a great work of art.

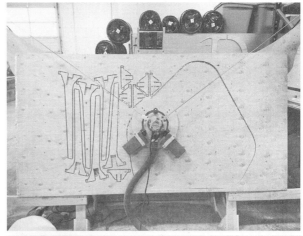

Figure 12.14   Maslow CNC machine. (CC BY-SA) https://www.maslowcnc.com/

Applications

With the three tools discussed in this chapter, you can make just about any physical object or product. 3D printers can be used to make everything from toys for your kids (Peteresen et al., 2017) to adaptive aids for arthritis sufferers (Gallup et al., 2018). 3D-printed toys generally save more than 75 percent of the cost of commercially available true alternative toys and over 90 percent if recyclebot filament is used (Peteresen et al., 2017). For example, for the travel chess set shown in Figure 12.15a, you would save about 90 percent if you printed it yourself compared with purchasing it on Amazon. With the 3D printer, you don't have to choose one particular chess set, but you could have dozens to pick from based on free online designs, as seen in a single page of results on Yeggi (Figure 12.15b). There are a lot more pages than this even for chess sets. Open-source toy designs can get quite complicated, but they are easy to print and fun to play with, as seen in Figure 12.16. Toys are generally mass produced from plastic and thus super-cheap,

Figure 12.15 (a) Visual comparison of open-source 3D-printed chess set and commercial mass-manufactured equivalents, along with costs and percent savings; (b) screenshot of Yeggi showing free online 3D-printable chess set designs. (CC-BY) https://www.mdpi.com/2227-7080/5/3/45/htm

Figure 12.16   Playing with 3D-printed toys, many articulated (with internal moving parts), printed in place.

yet open-source desktop 3D printers can make custom toys for far less. For other products, such as anything to do with medicine, that have huge markups, the savings are even greater. For example, the adaptive aids savings were even more than 94 percent for a wide range of the products seen in Figure 12.17 (Gallup et al., 2018). There are now millions of open-source 3D-printable designs housed in dozens of repositories (see reprap.org/wiki/Printable_part_sources).

A laser cutter can be used to make everything from intricate wedding invitations (Figure 12.18) to any of thousands of free designs that can be cut from solid materials like those shown in Figure 12.19. Sources for these tens of thousands of designs are summarized in Table 12.1. Have fun digital making in the comfort of your own home!

Figure 12.17   An assortment of 3D-printable open-source adaptive aids.

Figure 12.18   Elegant card with laser-cut gold detail by Freepick. (CC BY) https://www .freepik.com/free-vector/elegant-card -with-laser-cut-gold-detail_1163689 .htm

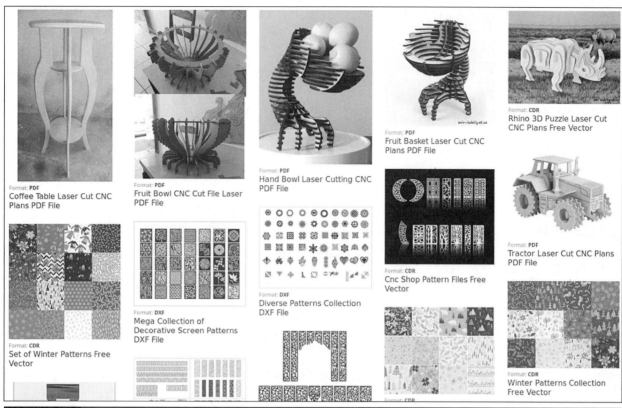

Figure 12.19 Screenshot of a few of the tens of thousands of free design files available on https://3axis.co/

Table 12.1 Free and Open-Source CNC Designs

CNC Plans Website	URL
3 Axis	3axis.co
CNC G-code	cncgcode.weebly.com
CNC Art Club	cncart.club
DXF for CNC	www.dxfforcnc.com/collections/free-dxf-files
Design Shop	www.designshop.com/search/free-cnc-patterns.html
DXF 1	dxf1.com
Free Patterns	www.freepatternsarea.com
Love SVG	lovesvg.com/free-svg-cut-files
My DXF	mydxf.blogspot.com
Obrary	obrary.com/collections/open-designs
Ponoko	www.ponoko.com/showroom/product-plans/free
Scan to CAD	www.scan2cad.com/free-downloads/dxf
Sign Torch	www.signtorch.com/store/Free-Vector-DXF-Art-Samples
Vision Engravers	www.visionengravers.com/support/vision-graphics-download.html

References

Arnott R. 2008. *The RepRap Project: Open Source Meets 3D Printing* (Computer and Information Science Seminar Series). University of Otago Library, Dunedin, New Zealand.

Bowyer A. 2004. Wealth without money. RepRap Wiki. http://reprap.org/wiki/Wealth_Without_Money

Bradshaw S, Bowyer A, Haufe P. 2010. The intellectual property implications of low-cost 3D printing. *ScriptEd* 7(1):1–27. https://doi.org/10.2966/scrip.070110.5

Cano J. 2011. The Cambrian explosion of popular 3D printing. *International Journal of Artificial Intelligence and Interactive Multimedia* 1(4):30–32.

Gallup N, Bow J, Pearce J. 2018. Economic potential for distributed manufacturing of adaptive aids for arthritis patients in the US. *Geriatrics* 3(4):89. https://doi.org/10.3390/geriatrics3040089

Gershenfeld N. 2005. *Fab: The Coming Revolution on Your Desktop—From Personal Computers to Personal Fabrication.* Basic Books, New York.

Hodgson G. 2013. Interview with Adrian Bowyer. *RepRap Magazine* 1: 8–14.

Jones R, Haufe P, Sells E. 2011. RepRap: the replicating rapid prototype. *Robotica* 29(1):177–191.

Pearce JM, Blair CM, Laciak KJ, et al. 2010. 3-D printing of open source appropriate technologies for self-directed sustainable development. *Journal of Sustainable Development* 3(4):17–29.

Petersen E, Kidd R, Pearce J. 2017. Impact of DIY home manufacturing with 3D printing on the toy and game market. *Technologies* 5(3):45. https://doi.org/10.3390/technologies5030045

Petersen EE, Pearce JM. 2017. Emergence of home manufacturing in the developed world: Return on investment for open-source 3-D printers. *Technologies* 5(1):7. https://dx.doi.org/10.3390/technologies5010007

Sells E, Bailard S, Smith Z, Bowyer A. 2007. RepRap: The replicating rapid prototype maximizing customizability by breeding the means of production. *In Handbook of Research in Mass Customization and Personalization* (in 2 Volumes) 2010 (pp. 568–580).

Weinberg M. 2010. It Will Be Awesome If They Don't Screw It Up. http://nlc1.nlc.state.ne.us/epubs/creativecommons/3DPrintingPaperPublicKnowledge.pdf

Wittbrodt B, Glover A, Laureto J, et al. 2013. Life-cycle economic analysis of distributed manufacturing with open-source 3-D printers. *Mechatronics* 23(6):713–726.

CHAPTER 13

Making and Sharing Scientific Equipment

Poor Scientists, Rich Open-Source Science

As you might suspect, scientists are also interested in the power of open-source technologies to save them money. Very few scientists have all the equipment they want to do their experiments. You may even want to do an experiment yourself and do not have the money for commercial scientific tools. Even the wealthiest labs at the largest companies and most prestigious universities don't have all the best scientific tools. This is largely due to the inflated prices of proprietary scientific equipment for experimental research (Pearce, 2014). When scientists don't have the tools they want, it slows the rate of scientific discovery.

If the relatively rich scientists can't afford all the tools they want, scientists at cash-strapped schools at lower levels do not have a chance. Nothing more than the high cost of scientific instruments limits access to exciting and engaging labs in both K–12 schools and colleges and universities (Gutnicki, 2010). This weakens recruitment into science, technology, engineering, and math (STEM) fields and results in a drain on scientific talent for the future. Historically, the scientific community had to choose one of two suboptimal paths to participate in state-of-the-art experimental research: (1) purchase absurdly high-cost patented tools or (2) develop equipment largely

from scratch in their own labs. You might think, "Scientists are smart—surely they can do option 2." You would be right, but it takes an enormous amount of time, and if every scientist has to build his or her own equipment all alone, he or she loses time to do his or her main experiments.

As you might suspect from reading the previous chapters in this book, open source may be the answer scientists have been looking for—and you would be right! Low-cost but often highly sophisticated and customized scientific equipment is being developed as free and open-source hardware (FOSH; Fisher and Gould, 2012; Pearce, 2012). Similar to what is seen in complex free and open-source software (FOSS) development (DiBona et al., 1999; Raymond, 1999; Soderberg, 2008), FOSH leads to improved product innovation in a wide range of scientific fields (Fisher and Gould, 2012; Hienerth et al., 2014; Pearce, 2014). The use of this open-source paradigm has become valuable for such diverse applications as obtaining costly clinical data for medical research (Dunn et al., 2012) and laboratory equipment. The open-source paradigm is now enabling the creation of FOSH for science by combining 3D printing (see Chapter 12) with open-source microcontrollers (see Chapter 11) running FOSS (Pearce, 2012, 2014). Scientists have been busy. They have made hundreds of scientific tools that are freely accessible both to themselves and to interested amateurs. This trend is assisting scientific development in every field that it touches (Pearce, 2014). Just as we learned in Chapter 1, research engineers design, share, and build on one another's work to develop scientific tools (Harnett, 2011).

Examples of Open-Source Science Tools

Open-source microcontrollers, such as the Arduino we discussed in Chapter 11, are being used for a wide range of chemical educational tools (e.g., simple colorimeters, pH meters, automated titrators, data loggers, nanoliter volume-dispensing systems, and generic control devices for automated assays [Urban, 2014]). An open-source Python framework has been developed for the Arduino that offers even more flexibility for applications (e.g., high-voltage power supplies; pressure and mass flow controllers; syringe pumps; two-, three-, and multiposition valves; and data-recording systems [Koenka et al., 2014]). Open-source microcontrollers can also be used for more sophisticated and targeted applications such as radial stretching systems with force sensors (Schausberger et al., 2015), saving hundreds of dollars. A robot-assisted mass spectrometry assay platform (Chiu and Urban, 2014) and an automated peptide synthesizer (Gali, 2017) have also been developed, saving over $25,000. Maybe one of these tools sparks your interest. Even if you have no interest at all in science, however, all these new open-source goodies may be of use to you. You could use many of the scientific tools to help you in your everyday life. For example, if you are going away for a few days, you could use an Arduino-controlled automatic feeder originally designed for animal experiments (Oh et al., 2017) to feed your pets. Other groups are developing open-source electronics to make Internet of Things (IoT) meter devices for smart and energy-efficient buildings (Pocero et al.,

2017) and sensor and computational platforms for smart cities (Jiang and Claudel, 2017), which could help reduce your electric and tax bills, respectively.

As we discussed in Chapter 12, open-source electronics drive OS 3D printers such as the RepRap (Wittbrodt et al., 2013; Anzalone et al., 2015), which are being used to manufacture high-quality scientific tools. Because the resulting complexity of manufacturing designs is free, it is just as easy to replicate an inexpensive test tube rack as it is to make an $850 magnetic rack. Just to be clear, a magnetic test tube rack is only a small piece of plastic that holds test tubes next to magnets. It is useful for separating out magnetic particles from a solution. It sounds really simple, and yes, you could absolutely just tape magnets to a normal test tube rack, but the real things were being sold for nearly $1,000. Researchers at the University of Washington got angry with these prices and designed their own 3D-printable magnetic test tube rack (Acadey, 2013). Because the magnets are available for only $6 each, it is possible to economically justify the purchase of a $500 open-source RepRap 3D printer for a lab by downloading, printing, and avoiding the cost of a single standard commercial magnetic rack!

The 3D printer can then be used to make a long list of progressively more sophisticated and costly tools (Pearce, 2014; Baden et al., 2015) that I have attempted to catalog on the webpage for the Open Source Lab (www.appropedia.org/Open-source_Lab) (Figure 13.1). They all have one thing in common with the consumer 3D printables we saw in Chapter 12: They cost less! Simply replacing the commercial centerpiece of an analytical ultracentrifuge with a free 3D-printed version can save scientists $1,000 (Desai et al., 2016). Similarly, an adjustable micropipette offsets a $1,000 lab purchase (Brennan et al., 2017). In addition, sharing digital designs and 3D printers can be used to

Open-source scientific hardware collections and resources

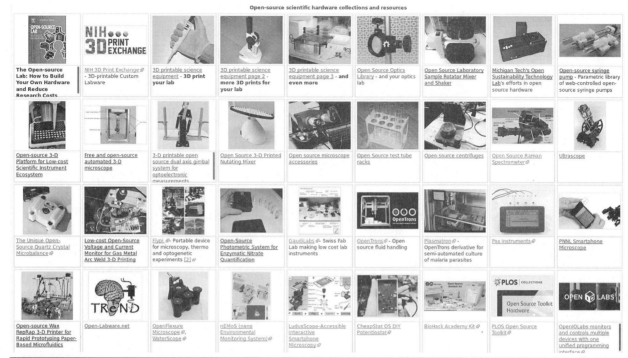

Figure 13.1 A screenshot of the Open Source Lab catalog of open-source scientific tools available on Appropedia. (CC BY-SA) https://www.appropedia.org/Open-source_Lab

attempt new experiments with novel plant tissue culture systems (Shukla et al., 2017), chemical reactionware (Symes et al., 2012), or polymer laser welders for heat exchangers (Arie et al., 2017). This latter device (Figure 13.2) is in my lab and has saved us literally thousands of dollars, because every time we turn it on to make a new type of heat exchanger, it saves us having to purchase laser weld time at a commercial facility that costs hundreds of dollars per prototype. Heat exchangers are one of those expensive things that you might just need a replacement for, whether you need one for your home heating, ventilation, and air-conditioning (HVAC) system or you are building a home water pasteurizer (Denkenberger and Pearce, 2018).

The most powerful and expensive open-source scientific tools combine 3D printing (see Chapter 12) and open-source electronics (see Chapter 11). For example, several approaches have been shown to decrease the cost of microfluidics

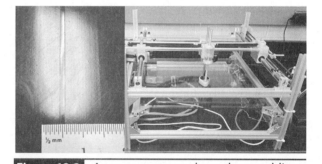

Figure 13.2 An open-source polymer laser welding system made mostly with 3D-printed parts and controlled with the same open-source electronics that operates a RepRap 3D printer (Laureto et al., 2017). This welding system is used to make novel heat exchangers by additively manufacturing 2D polymer sheets into complex 3D structures. (CC BY-SA) https://www.appropedia.org/Open_Source_Laser_Polymer_Welding_System:_Design_and_Characterization_of_Linear_Low-Density_Polyethylene_Multilayer_Welds

platforms, saving researchers $2,000 or more (Pearce et al., 2016; Tothill et al., 2017). In addition, a handheld portable open-source colorimeter (which is discussed in more detail in Chapter 14) can be built to do chemical oxygen demand (COD) measurements for under $50, replacing a proprietary handheld tool that cost more than $2,000 (Anzalone et al., 2013). This relatively simple device can be extended to do nephelometry (measuring the concentration of suspended particulates in liquids) by spending a few more dollars to buy more light-emitting diodes (LEDs), thus saving another $600 to $2,000 (Wijnen et al., 2014), or it can be adapted to make nitrate measurements (Wittbrodt et al., 2015). Similarly, an open-source quartz crystal microbalance that we use in my lab all the time saves us more than $2,000 without compromising quality (Mista et al., 2016). Moreover, a single automated device such as a filter wheel changer can be built in a day for $50, replacing inferior commercial tools that cost $2,500 (Pearce, 2012).

Even better, such specialty tools often have huge lead times, so with 3D printing you can get the tool you want without the wait. Not only does a 3D-printable open-source optics library already exist (Zhang et al., 2013), but it has been expanded (Gopalakrishnan and Gühr, 2015; Salazar-Serrano et al., 2017). In addition, scientists are pushing ever more complex tools, such as an open mesoscope (Gualda et al., 2014); an open, high-content fluorescence lifetime microscope (Barber et al., 2013; Görlitz et al., 2017); a one-piece microscope with flexure translation (Sharkey et al., 2016); an automated 3D microscope (which saves several thousand dollars to more than $10,000 [Wijnen et al., 2016]); and even such top-end equipment as an $800 open-source microscope that replaces an $80,000 conventional equivalent (Open Lab Tools 2014).

There is no limit to the savings, just as there is no limit to the sophistication. Now DNA nanotechnology lab gel scanners, horizontal polyacrylamide gel electrophoresis (PAGE) gel molds, and homogenizers for generating DNA-coated particles can all be fabricated using an open-source approach for up to a 90 percent savings (Damase et al., 2015). Similarly, there is a long list of open-source money-saving tools for biology labs (e.g., microscopes, thermocyclers for polymerase chain reaction [PCR] testing, centrifuges, hotplates, magnetic stirrers, waveform generators, electroencephalographs [EEGs], and Skinner boxes, to name a few examples [Baden et al., 2015]).

The open-source 3D printers being driven by open-source electronics can even become scientific tools themselves. For example, a 3D printer can be used to make the thin silica gel layers used in planar chromatography (Fichou and Morlock, 2017). This type of control of a 3D printer is particularly easy if the device uses Franklin, an open-source 3D motion-control software suite (Wijnen et al., 2016). Franklin has been used to make an automated mapping four-point probe (Chandra et al., 2017), as shown in Figure 13.3. This device normally would cost about a quarter of a million dollars and be seen only in the most state-of-the-art semiconductor labs (e.g., you might run across one at Intel). Our version can be built for a few thousand dollars, saving more than the cost of a decent home in most parts of the United States.

You can build some of the most sophisticated 3D scientific platforms yourself, such as the device shown in Figure 13.4 (Zhang et al., 2016). This open-source 3D scientific platform was designed for automatic stirring, measuring, and probing in the laboratory, as well as automated fluid handling, and it even does shaking and mixing (thereby taking the place of dedicated open-source tools such as the simple mixer

Figure 13.3 Open-source heated mapping four-point probe (Chandra et al., 2017) used to measure the conductivity of the surface of thin films and other materials. (CC BY-SA) https://www.appropedia.org/Open-Source_Automated_Mapping_Four-Point_Probe

shown in Figure 13.5 [Dhankani and Pearce, 2017]). You could use this system as a baker bot. The automated stirring can be used for brownie batter just as easily as it could be used to stir chemicals in a lab. Shake and bake might be more fun if you mount it on the platform. You can even use the shaking function to take the labor out of making homemade ice cream. You can get the following recipe online, as we discussed in Chapter 2, and use this device to make some tasty strawberry ice cream: double bag of ice plus salt in the outer bag, with an inner bag containing 2 cups of whole milk, 2 cups of heavy cream, 1 cup of sugar, 2 teaspoons of vanilla extract, and 2 cups of mashed strawberries.

You also might find some of the science features a natural fit for a hobby. Do you like taking pictures of flowers? If you put an inexpensive USB microscope on the platform, you can use it as a sophisticated 3D microscope (Wijnen et al., 2016). With free software called *imageJ* (imagej.nih.gov/ij), these microscopes can do automated stitching (where you take a bunch of images of a surface and make a big one) or stacking (where you take pictures at different focal lengths and combine them). This technique is demonstrated on a flower in the insets of Figure 13.4. The top-right inset shows the flower

Figure 13.4 Open-source 3D platform for low-cost scientific instrument ecosystem. Mounted on the end effector is a USB microscope (Wijnen et al., 2016). In the background is a standard fused-filament print head (gray), which was used to print out the syringe pump used for fluid handling shown in the foreground, as well as the various scientific and engineering tools shown on the left in yellow plastic (e.g., PCB mill, glass stir rod holder, etc.). Full details are available in Zhang et al., 2016. (CC BY-SA) https://www.appropedia.org/Open-source_3-D_Platform_for_Low-cost_Scientific_Instrument_Ecosystem

Figure 13.5 An open-source 3D-printed sample holder and laboratory sample rotator/mixer and shaker controlled with an open-source Adafruit Pro trinket microcontroller (Dhankani and Pearce, 2017). (CC BY-SA) https://www.appropedia.org/Open_Source_Laboratory_Sample_Rotator_Mixer_and_Shaker

bottom in focus, the middle inset shows the top, and the bigger inset shows what happens when you stack all the images together: a clear, crisp, better picture!

Open source not only radically reduces the cost of doing science, but it can also help train future scientists (Pearce, 2013). An entire university physics classroom of optics setups can be 3D printed for $500 using predesigned components from the open-source optics library on an open-source 3D printer, replacing $15,000 of commercial equipment (Zhang et al., 2013). This would save over $66 million if scaled only to the basic physics laboratories in degree-granting institutions in the United States (U.S. Department of Education, 2013).

Better yet, why should university students get to have all the fun? If we made a set for every secondary school in the United States, it would save over $500 million! (And it would provide high school students with high-quality scientific equipment to improve their experiences in science classes.) There is no question that huge sums of money could be saved, and the return on investment for getting schools to invest in open science would be huge (Pearce, 2016)—it could reach more than 1,000 percent!

Some of the funding agencies have received the memo on the power of open source. For example, the National Institutes of Health (NIH), the largest American scientific funder, not only demands that everything it funds be made open access, but it has also created the NIH 3D Print Exchange. The exchange has a bunch of cool designs of things such as protein molecules for learning, but it also has a section on custom labware (3dprint.nih.gov; Coakley et al., 2014; Coakley and Hurt, 2016). Most recently it was used to house 3D printable designs that were approved for clinical or community use during the COVID-19 pandemic.

Next Steps

It is now well established that knowledge sharing via networked science has incredible power (Salter and Martin, 2001; Lang, 2011; Woelfle et al., 2011). "Crowd science" (Young, 2010), "citizen science" (Wiggins and Crowston, 2011), "networked science" (Nielsen, 2011), and "massively collaborative science" (Franzoni and Sauermann, 2014) all will benefit from low-cost scientific hardware, enabling their practitioners to go far beyond collaborating on computer simulations alone. I will talk about how this works in Chapter 14.

We also know that it will be good for education: Urban (2014) shows how open-source electronics can aid in chemical education; Hill and Ciccarelli (2013) show how FOSH can help teach even young students programming skills; Zhang et al. (2013) show how FOSH can be used to reduce costs in physics education; and Kentzer et al. (2011) show how it can be used to teach mechatronic engineering. This scaled replication provides savings generally between 90 and 99 percent of the commercial proprietary costs of scientific tools (Pearce, 2014).

This benefits all of science, which is a big deal. Why? Because improvements in science lead to improvements in technology, which will enhance every aspect of the economy—that is, your life (Salter and Martin, 2001). Historically, the secondary benefits of funding science were on the order of 20 to 70 percent (Salter and Martin, 2001), but a study on the return on investment (ROI) of an open-source syringe pump (Wijnen et al., 2014) found that the ROI ranged from hundreds to thousands of percent after only a few months (Pearce, 2016). It is clear that we are only scratching the surface of open science and the wonders that it will bring. To learn how you can be a part of it, keep reading!

References

Acadey. 2013. 96-well plate/0.2-mL strip tube magnet rack. http://www.thingiverse.com/thing:79430

Anzalone GC, Glover AG, Pearce JM. 2013. Open-source colorimeter. *Sensors* 13(4): 5338–5346.

Anzalone GC, Wijnen B, Pearce JM. 2015. Multi-material additive and subtractive prosumer digital fabrication with a free and open-source convertible delta RepRap 3D printer. *Rapid Prototyping Journal* 21(5): 506–519.

Arie MA, Shooshtari AH, Tiwari R, et al. 2017. Experimental characterization of heat transfer in an additively manufactured polymer heat exchanger. *Applied Thermal Engineering* 113:575–584.

Baden T, Chagas AM, Gage G, et al. 2015. Open labware: 3D printing your own lab equipment. *PLOS Biology* 13(3). https://doi.org/DOI: 10.1371/journal.pbio.1002086

Barber PR, Tullis IDC, Pierce GP, et al. 2013. The Gray Institute "open" high-content, fluorescence lifetime microscopes. *Journal of Microscopy* 251(2):154–167.

Brennan M, Bokhari F, Eddington D. 2017. Open design 3D-printable adjustable micropipette that meets ISO standard for accuracy. *Micromachines* 9(4):191–195.

Chandra H, Allen SW, Oberloier SW, et al. 2017. Open-source automated mapping four-point probe. *Materials* 10(2):110. https://doi.org/doi:10.3390/ma10020110

Chiu SH, Urban PL. 2015. Robotics-assisted mass spectrometry assay platform enabled by open-source electronics. *Biosensors and Bioelectronics* 64:260–268.

Coakley MF, Hurt DE. 2016. 3D printing in the laboratory: Maximize time and funds with customized and open-source labware. *Journal of Laboratory Automation* 21(4):489–495.

Coakley MF, Hurt DE, Weber N, et al. 2014. The NIH 3D print exchange: a public resource for bioscientific and biomedical 3D prints. *3D Printing and Additive Manufacturing* 1(3):137–140.

Damase TR, Stephens D, Spencer A, Allen PB. 2015. Open source and DIY hardware for DNA nanotechnology labs. *Journal of Biological Methods* 2(3):e24.

Denkenberger DC, Pearce JM. 2018. Design optimization of polymer heat exchanger for automated household-scale solar water pasteurizer. *Designs* 2(2):11. https://doi.org/10.3390/designs2020011

Desai A, Krynitsky J, Pohida TJ, et al. 2016. 3D-printing for analytical ultracentrifugation. *PloS ONE* 11(8):e0155201.

Dhankani KC, Pearce JM. 2017. Open source laboratory sample rotator mixer and shaker. *HardwareX* 1:1–12.

DiBona C, Ockman S, Stone M. 1999. *Open Sources: Voices from the Open Source Revolution*, 1st ed. O'Reilly, Sebastopol CA.

Dunn AG, Day RO, Mandl KD, Coiera E. 2012. Learning from hackers: Open-source clinical trials. *Science Translational Medicine* 4(132):132cm5. DOI: 10.1126/scitranslmed.3003682

Fichou D, Morlock GE. 2017. Open-source-based 3D printing of thin silica gel layers in planar chromatography. *Analytical Chemistry* 89(3):2116–2122.

Fisher D, Gould P. 2012. Open-source hardware is a low-cost alternative for scientific instrumentation and research. *Modern Instrumentation* 1(2):8–20.

Franzoni C, Sauermann H. 2014. Crowd science: The organization of scientific research in open collaborative projects. *Research Policy* 43(1):1–20.

Gali H. 2017. An open-source automated peptide synthesizer based on Arduino and Python. *SLAS Technology: Translating Life Sciences Innovation 22*(5), 493-499.

Gopalakrishnan M, Gühr M. 2015. A low-cost mirror mount control system for optics setups. *American Journal of Physics* 83(2):186–190.

Görlitz F, Kelly DJ, Warren SC, et al. 2017. Open source high content analysis utilizing automated fluorescence lifetime imaging microscopy. *Journal of Visualized Experiments* (119), p.e55119.

Gualda E, Moreno N, Tomancak P, Martins GG. 2014. Going "open" with mesoscopy: A new dimension on multi-view imaging. *Protoplasma 251*(2), 363-372.

Gutnicki J. 2010. The evolution of teaching science. *Innovative Educator*. http://theinnovativeeducator.blogspot.com/2010/02/evolution-of-teaching-science.html

Harnett C. 2011. Open source hardware for instrumentation and measurement. *IEEE Instrumentation and Measurement Magazine 14*(3), 34–38. DOI: 10.1109/MIM.2011.5773535

Hienerth C, von Hippel E, Berg Jensen M. 2014. User community vs. producer innovation development efficiency: A first empirical study. *Research Policy* 43(1):190–201.

Hill L, Ciccarelli S. 2013. Using a low-cost open source hardware development platform in teaching young students programming skills. In *Proceedings of the 13th Annual ACM SIGITE Conference on Information Technology Education*, pp. 63–68. https://doi.org/10.1145/2512276.2512289

Jiang J, Claudel C. 2017. A high performance, low power computational platform for complex sensing operations in smart cities. *HardwareX* 1:22–37.

Kentzer J, Koch B, Thiim M, et al. 2011. An open source hardware-based mechatronics project: The replicating rapid 3D printer. In *Proceedings of the 4th International Conference on Mechatronics (ICOM)*. (pp. 1-8). IEEE, Piscataway, NJ, pp. 1–8.

Koenka IJ, Sáiz J, Hauser PC. 2014. Instrumentino: An open-source modular Python framework for controlling Arduino-based experimental instruments. *Computer Physics Communications* 185(10):2724–2729.

Lang T. 2011. Advancing global health research through digital technology and sharing data. *Science* 331:714–717.

Laureto JJ, Dessiatoun SV, Ohadi MM, Pearce JM. 2016. Open source laser polymer welding system: Design and characterization of linear low-density polyethylene multilayer welds. *Machines* 4(3):14. https://doi.org/doi:10.3390/machines4030014

Mista C, Zalazar M, Peñalva A, et al. 2016. Open source quartz crystal microbalance with dissipation monitoring. *Journal of Physics: Conference Series* 705(1):012008.

Nielsen M. 2011. *Reinventing Discovery: The New Era of Networked Science*. Princeton University Press, Princeton, NJ.

Oh J, Hofer R, Fitch WT. 2017. An open source automatic feeder for animal experiments. *HardwareX* 1:13–21.

Open Lab Tools. 2014. The OpenLabTools project. University of Cambridge. http://openlabtools.eng.cam.ac.uk/

Pearce JM. 2012. Building research equipment with free, open-source hardware. *Science* 337(6100):1303–1304.

Pearce JM. 2013. Open-source hardware for research and education. *Physics Today* 66(11), 8. https://doi.org/doi:10.1063/pt.3.2160

Pearce JM. 2014. *Open-Source Lab: How to Build Your Own Hardware and Reduce Research Costs*. Elsevier, New York.

Pearce JM. 2016. Return on investment for open source hardware development. *Science and Public Policy* 43(2):192–195.

Pearce JM, Anzalone NC, Heldt CL. 2016. Open-source wax reprap 3D printer for rapid prototyping paper-based microfluidics. *Journal of Laboratory Automation* 21(4): 510–516.

Pocero L, Amaxilatis D, Mylonas G, Chatzigiannakis I. 2017. Open source IoT meter devices for smart and energy-efficient school buildings. *HardwareX* 1:54–67.

Raymond E. 1999. The cathedral and the bazaar. *Knowledge, Technology and Policy* 12(3):23–49.

Salazar-Serrano LJ, Torres JP, Valencia A. 2017. A 3D printed toolbox for opto-mechanical components. *PloS ONE* 12(1), e0169832.

Salter AJ, Martin BR. 2011. The economic benefits of publicly funded basic research: A critical review, *Research Policy* 30:509–532.

Schausberger SE, Kaltseis R, Drack M, et al. 2015. Cost-efficient open source desktop size radial stretching system with force sensor. *IEEE Access* 3:556–561.

Sharkey JP, Foo DC, Kabla A, et al. 2016. A one-piece 3D printed flexure translation stage for open-source microscopy. *Review of Scientific Instruments* 87(2):025104.

Shukla MR, Singh AS, Piunno K, et al. 2017. Application of 3D printing to prototype and develop novel plant tissue culture systems. *Plant Methods* 13(1):6.

Soderberg J. 2008. *Hacking Capitalism: The Free and Open Source Software Movement.* Routledge, London.

Symes MD, Kitson PJ, Yan J, et al. 2012. Integrated 3D-printed reactionware for chemical synthesis and analysis. *Nature Chemistry* 4(5):349–354.

Tothill AM, Partridge M, James SW, Tatam RP. 2017. Fabrication and optimisation of a fused filament 3D-printed microfluidic platform. *Journal of Micromechanics and Microengineering* 27(3):035018.

Urban PL. 2014. Open-source electronics as a technological aid in chemical education. *Journal of Chemical Education* 91 (5):751–752.

U.S. Department of Education. 2013. *Digest of Education Statistics 2013*. National Center for Education Statistics. http://nces.ed.gov/programs/digest/2013menu_tables.asp

Wiggins A, Crowston K. 2011. From conservation to crowdsourcing: A typology of citizen science. Paper presented at the 44th Hawaii International Conference on Systems Sciences (HICSS). IEEE. Hawaii, USA.

Wijnen B, Anzalone GC, Haselhuhn AS, et al. 2016. Free and open-source control software for 3D motion and processing. *Journal of Open Research Software* 4(1):e2. https://doi.org/10.5334/jors.78

Wijnen B, Anzalone GC, Pearce JM. 2014. Open-source mobile water quality testing platform. *Journal of Water Sanitation and Hygiene for Development* 4(3):532–537.

Wijnen B, Hunt EJ, Anzalone GC, Pearce JM. 2014. Open-source syringe pump library. *PloS ONE* 9(9), e107216.

Wijnen B, Petersen EE, Hunt EJ, Pearce JM. 2016. Free and open-source automated 3D microscope. *Journal of Microscopy* 264(2):238–246.

Wittbrodt BT, Glover AG, Laureto J, et al. 2013. Life-cycle economic analysis of distributed manufacturing with open-source 3D printers. *Mechatronics* 23(6):713–726.

Wittbrodt BT, Squires DA, Walbeck J, et al. 2015. Open-source photometric system for enzymatic nitrate quantification. *PloS ONE* 10(8):e0134989.

Woelfle M, Olliaro P, Todd MH. 2011. Open science is a research accelerator. *Nature Chemistry* 3(10), 745–748.

Young J. 2010. Crowd science reaches new heights. *Chronicle of Higher Education*, May 28. http://chronicle.com/article/The-Rise-of-Crowd-Science/65707/

Zhang C, Anzalone NC, Faria RP, Pearce JM. 2013. Open-source 3D-printable optics equipment. *PLoS ONE* 8(3):e59840. https://doi.org/10.1371/journal.pone.0059840

Zhang C, Wijnen B, Pearce JM. 2016. Open-source 3D platform for low-cost scientific instrument ecosystem. *Journal of Laboratory Automation* 21(4):517–525.

CHAPTER 14

Making and Sharing Data as a Citizen Scientist

Introduction to Citizen Science

As we talked about in Chapter 13, scientists and makers who are sharing scientific tools enable everyone to be a scientist. This is helping a growing trend called *citizen science*. Citizen science (sometimes called *crowdsourced science* or *civic science*) is scientific research conducted, in whole or in part, by amateur scientists. You can become a citizen scientist simply by wanting to be one. Citizen science is just scientific research done by members of the general public, although it is often done in collaboration or under the direction of professional scientists and fancy scientific institutions. Citizen science can be really useful, and somewhat amazingly, volunteer citizen scientists can provide data that are in general of reliably high quality, on par with those produced by professionals (Kosmala et al., 2016).

As a working scientist, I can tell you from first-hand experience that it is a really good feeling to discover something about our world that no one else knows and to share that information with everyone else to try to help the common good. To be a professional scientist involves being dedicated to long, hard, and expensive training, which is a pretty big barrier for most people to enjoying my profession. Now, with the advent of citizen science, however, literally everyone can donate a small amount of their time to participate in the scientific endeavor and make a concrete contribution to the body of knowledge for humanity. There are hundreds of projects you can become involved with, which I will discuss in the next section. My advice is to try a bunch of them to see what you like and only commit to the ones you are really interested in personally. In addition, if you have children doing citizen science attached to whatever they are learning about, it becomes much more interesting and easier to learn. It is good to be able to learn answers given in a textbook, but it is much more exciting to help identify real wildlife or discover new galaxies.

Easy First Steps into Citizen Science

Citizen science may appear to be a bit intimidating at first, but you probably already do things—like posting selfies on the web—that can be easily adapted to participate in citizen science. Take, for example, Project Soothe (www.projectsoothe.com). It is all about helping us soothe ourselves. Research and psychological therapy have shown that if you have the ability to soothe yourself at times of distress, it will help you to stay well. But how do people actually do this in everyday life? It turns out that scientists don't have a good handle on that yet. Project Soothe is helping them get a better understanding by having you send them photos that make you feel soothed. From Chapter 3, you already know how to use open-source tools and licenses to deal with photos. Project Soothe asks that you post your soothing images on its website gallery, and then it collects feedback from viewers about whether your images make them feel soothed as well. All the images from all the citizen scientists will then be combined into a large bank of photos for use in future research and psychological therapies.

Zooniverse

You can join Zooniverse (www.zooniverse.org), which is the world's largest group (more than a million volunteers) for research powered by citizen scientists just like you. Currently, Zooniverse has more than 100 active projects, and volunteers have contributed data to more than 2,500 published studies (as of 2020). You can join hundreds of thousands of volunteers

from around the world who come together to assist professional researchers while learning a little science for fun at the same time. Zooniverse projects are not passive—they require your active participation to complete research tasks. Projects have been drawn from such disciplines as astronomy, ecology, cell biology, humanities, and climate science. Zooniverse has projects in nearly every area of interest, including, for example, the following:

- Helping analyze wildlife photographs to track big cats, such as the leopard in Figure 14.1. A critical part of protecting big cats and their landscapes is documenting the presence and behavior of wild cats using automated cameras. Every year, Panthera's motion-activated cameras collect hundreds of thousands of wildlife images. With your help, the Panthera project can identify the animals, which enables tracking of

Figure 14.1   A screen shot of a citizen science project to analyze wildlife photos to help Panthera protect big cats. https://www.zooniverse.org/projects/panthera-research/camera-catalogue/classify

wild cat population trends over time and determination of the best conservation actions to better protect them.

- Helping the Smithsonian Institution better understand climate change with the Fossil Atmospheres Project. You can help to create a record of how the atmosphere has changed through time by calculating the ratio of two different types of leaf cells (stomatal and epidermal) for many leaves from the present and from the geologic past. We know that the composition of the Earth's atmosphere has changed over time and that it is changing now. It is important for us to understand what effect climate change might have on life on our planet in the long term, and your digital microscopy work can help to do it.

- Likewise, helping to sort out the complexity of sunspots to better understand the sun.

- Do you have a good ear? Maybe you could help characterize bat calls to better understand bats.

- Helping scientists better understand disease and support human health with the project Science Scribbler: Virus Factory. This project is currently trying to understand how viruses multiply inside cells, which seems particularly urgent today.

- Helping the physics community search for unknown exotic particles.

The Zooniverse also has projects that are outside the hard sciences, such as in history (e.g., annotate war diaries from World War I) and the arts (e.g., transcribe documents from long-dead artists or help understand Shakespeare better by learning about his contemporaries).

Citizen Science for Uncle Sam

CitizenScience.gov is the official government website for crowdsourcing and citizen science across all the scientific research sponsored by the U.S. government. It has a catalog of dozens of federally supported citizen science projects that you can use to get involved. These projects can be fun afternoons in your backyard or can be deadly serious. For an example of pure fun, Butterflies and Moths of North America (BAMONA; www.butterfliesandmoths.org) is collecting, storing, and sharing species information and occurrence data. You can participate by taking and submitting photographs of butterflies, moths, and caterpillars in your backyard. The data entry is simple, as shown in Figure 14.2, and your results are verified by professionals. The BAMONA project is based on work previously supported by the U.S. Geological Survey.

Figure 14.2   A screen shot of one of millions of verified moth and butterfly sightings. This one is a Baltimore Checkerspot used by the BAMONA project. http://www.butterfliesandmoths.org/

Not all insects are pure fun. In fact, some insects are the exact opposite of fun, like the irritating mosquito responsible for outbreaks of transmitted pathogens such as West Nile virus, Eastern equine encephalitis, dengue, and chikungunya. To help battle back against the mosquitoes, you can join the Invasive Mosquito Project (www.citizenscience.us/imp). It is aimed at monitoring invasive container-inhabiting mosquito species across America. Your research will help determine where the native and invasive mosquito species are distributed across the United States and figure out who is at risk. This is a bit of a more hard-core citizen science project because it provides anyone interested with the opportunity for training to collect real data and contribute to a national mosquito species distribution study. This project not only gives you an opportunity to explore and collect data around your house, it also raises awareness of diseases that can be transmitted by mosquitoes and how you can protect yourself, your family, and your pets from illness.

You can also help medical scientists better understand people so they can help cure diseases you may have in the future. For example, EyeWire (eyewire.org) investigators are solving the mysteries of the brain with your help. Best of all, you do this by playing a game that is really a 3D neuroscience coloring book. This puzzle game that anyone can play without any experience in neuroscience helps EyeWire researchers map the retina. Eventually, EyeWire wants to map the human brain, but it is starting small. When you play the game, you are mapping the connections between retinal neurons, helping researchers understand how neurons process information. Past game players have already helped researchers understand how we can detect motion, and they hope that your help can lead to advances in blindness therapies and the development of retinal prosthetics. Not a bad outcome for playing a fun game!

Citizen science projects are much bigger than our bodies and backyards—they go all the way to space. NASA has been an early and aggressive user of citizen science. NASA sponsors NASA SOLVE (www.nasa.gov/solve), which is a one-stop shop to find opportunities to participate in challenges, prize competitions, and citizen science activities that help NASA's mission. For example, you can help NASA scientists map the moon with Cosmoquest (cosmoquest.org/x/science/moon). The Lunar Reconnaissance Orbiter provided the images necessary to map out features only a few meters across over the surface of the moon. Unfortunately, software can't do this mapping yet, and NASA doesn't have the human resources needed to map out the entire moon. NASA needs to collect a diverse enough data set that it can train machine-learning algorithms to find craters in different kinds of soils and under different lighting conditions. To make sure that nothing is missed, NASA scientists have overlaps, and each area is looked at using different magnifications. The scientists will show you some images from a ½ to 1 square kilometer in size. Your efforts will train algorithms and speed up the process. This is a chance to do some good for NASA while you look at some super-close-up images of the moon, as in Figure 14.3. This is not something you probably want to do for a long time, but NASA has set it up so you can listen and learn about the moon while you flex your citizen science mental muscles.

Not all of the NASA projects are done on a computer. For example, if you live in the North and like to go outside at night, you can join the Aurorasaurus project (aurorasaurus.org). The goal for Aurorasaurus is to spot the aurora and report it. Your report will help others see it. Acting as a volunteer, you can use the website or Android apps to get alerts, help find real-time sightings from social media, and learn about the auroras.

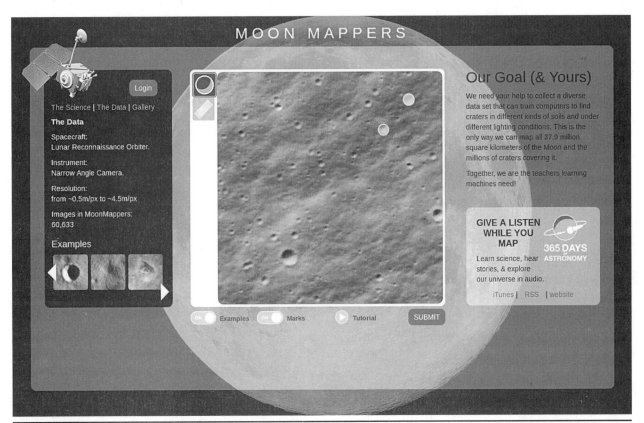

Figure 14.3 A screenshot of Moon Mappers, where I have identified two craters on the moon for NASA. https://cosmoquest.org/x/science/moon/

If you are really interested in space and have your own telescope, you could help NASA keep track of asteroids with Target Asteroids! (www.asteroidmission.org). You send your telescope images and data for asteroids that are important to the OSIRIS-REx asteroid sample-return mission and directly help our understanding of near-Earth asteroids. These combined observations enable scientists at NASA and at universities throughout the world to better understand the physical properties of asteroids for which little information is known, refine their orbits, and determine their parent asteroid families. If you don't own a telescope, NASA also runs annual month-long measuring campaigns for citizen scientists to provide images from large telescopes at dark-sky sites to a limited number of "measurers," who submit asteroid data to the Target Asteroids! program.

Meanwhile, NASA is also interested in helping us better understand our own planet with (again) your help. For example, landslides cause billions of dollars in infrastructural damage and thousands of deaths every year worldwide. As a society, we can try to prevent them and prevent the damage they cause from being so great. Data on past landslide events help guide us to make better future disaster prevention plans. Unfortunately, it is pretty hard to get data on all the landslides. We don't have a global picture of exactly when and where landslides occur. NASA is building the biggest open global landslide inventory to address this problem. The Cooperative Open Online Landslide Repository (COOLR) is an open platform where citizen scientists can share landslide reports. You can add data to COOLR using the Landslide

Reporter (landslides.nasa.gov/reporter) and then see all landslide data from COOLR with Landslide Viewer (landslides.nasa.gov/viewer).

When You Can't Trust the Government: Measure Your Own Risk

Although, in general, government scientists are trustworthy, there have been many attempts to either muzzle them, distort their findings, or cut their funding when they say politically uncomfortable things. In the popular media, this has been labeled a "war on science." Long before science was so politicized in the United States, however, there were specific technical areas where, for whatever reason, the American public had only been given an incomplete truth. One of those areas is the nuclear power industry, which is the only business incapable of covering its own liability insurance cost without a government guarantee (Pearce, 2009, 2012; Zelenika-Zovko and Pearce, 2011). The U.S. Nuclear Regulatory Commission (1983) found that the economic value of this liability cap was so large that it could be referred to as a subsidy. Calculating how much nuclear liability insurance is subsidized with the arbitrary liability caps is really hard, and there are a bunch of studies trying to figure it out. It should be easy, however, to figure out how much things cost after a disaster because postmortem calculations can be used to calculate insurance costs—or at least that is what I thought.

The earthquake and tsunami that struck Japan in March 2011 and caused nuclear radiation leakages in Fukushima presented a unique opportunity—a modern disaster in a technically advanced democratic and open society. Using openly available data, it should have been easy to calculate the insurance subsidy. We looked at the Fukushima (Daiichi) nuclear power plant disaster to find the true cost of the liability insurance needed to cover the damage based on news reports. This sounds easy, but the news reports all conflicted with one another. There were no retractions, no fixed values. The reported costs of the Fukushima nuclear disaster are somewhere between $20 billion and $525 billion (Laureto and Pearce, 2016). This is an absurd spread, but this makes the real insurance cost of the lifetime of electricity produced at the highly sophisticated Japanese nuclear plants calculated at between $0.22 and $5.78 per kilowatt-hour. Even the lower value is more than what most Americans pay for electricity. These values are far higher than the current insurance cost mandated by Japanese law of $0.01 per kilowatt-hour and even the total cost consumers pay for electricity in Japan. It wasn't just the real economic cost of nuclear power that was being kept from the Japanese people, it was also the risk.

What were those risks? A major impetus for the explosion of open-source Geiger counter projects (Pearce, 2013) was to help people in Japan measure the levels of radiation in their everyday life after Fukushima. Official announcements of nuclear accidents are viewed as unreliable by much of the public in both Japan and the United States, primarily because of conflicts of interest. The government is going to have to pick up the bill for any disaster because it backstops the industry's insurance. The government therefore has an incentive to downplay any disaster. During the Fukushima disaster, the *Japanese Times* reported that the Japanese public was extremely skeptical of official reports in part because government officials appeared to be actively preventing citizens from obtaining data (Brasor, 2012). Even in the United States, where we would hope that communication would be better, our government appeared reluctant to post online whatever levels it was monitoring as radiation from Fukushima hit the West Coast, and *The Washington Post* (2011) reported that its monitors crashed.

The response from citizen scientists in Japan was fast and furious. They crowdsourced Geiger counter radiation readings from across Japan using a collection of both open-source hardware and open-source software for the Japan Geigermap. This did not even need to be particularly complicated hardware hacking. There is already an Android app that can turn your cell phone into a tolerable radiation detector using the CMOS camera covered by a piece of tape (Benchoff, 2012). Even better, Softbank launched a smartphone that tracks radiation specifically for Japan that you can use in America as well. Crowdsourced radiation or pollution levels and similar kinds of activity may be the start of a movement to bring more people into science to act as a check on bad or dishonest sources of scientific information for the public (Pearce, 2012).

Do not be lulled into complacency and think this sort of thing could not happen in America. Consider my home state of Pennsylvania's Three Mile Island nuclear accident from decades ago. There is now a figurative ton of evidence that the releases were underreported to the public by officials by, at the very least, a factor of 2, and maybe even by a factor of 1,000. The official U.S. Nuclear Regulatory Commission (NRC) value is 10 MCi from the President's Commission on the Accident at Three Mile Island (1979). A follow-up study more than doubled that to 22 MCi at minimum (Thompson et al., 1995), whereas a careful look at the sum of the nuclear releases yields 36 MCi and estimates anywhere between 100 and 1,000 times the reported value (Gundersen, 2009). Will we ever know the truth? Maybe not, because there was not a fleet of distributed citizen science–operated open-source Geiger counters around Pennsylvania at the time. The bottom line, however, perhaps comes from the medical community: A raft of epidemiological studies points to a significant epidemic of cancer that is clearly related to the Three Mile Island release (Gur et al., 1983; Hatch et al., 1990, 1991; Wing et al., 1997, 2000; Talbott et al., 2000a, 2000b). Simply put, if the NRC values were accurate, then the health impacts around Three Mile Island would have been a lot less. If you live anywhere near a nuclear power plant (see Figure 14.4), it might be worth looking into an open-source Geiger counter project. Numerous open-source projects are available, such as the PiGi with a Raspberry Pi, and the MicroGeiger, Seeedstudio, PocketMagic, and Sparkfun Geiger counters.

U.S. Operating Commercial Nuclear Power Reactors

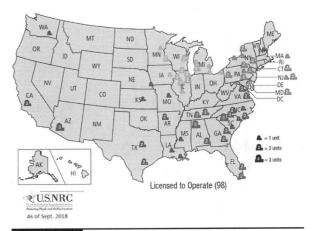

Figure 14.4 Map of the current 98 operating commercial nuclear reactors in the United States. Map compliments of U.S. Nuclear Regulatory Commission. (Public domain) https://www.nrc.gov/reactors/operating/map-power-reactors.html

How Companies Can Harness Open Source to Support Citizen Scientists

You can have too much of a good thing. Take nutrients, for instance. Nutrients can help your body grow or help crops on a farm to grow. Too much, however, can be poison. Consider nitrates and nitrites, which are used as fertilizers. When we treat sewage, fertilize crops, and have large animal farms, we end up with too many nutrients in virtually all waters of the Earth (Canfield and Glazer, 2010). According to the U.S. Environmental Protection Agency (EPA, 2015), among all environmental problems in America, excess nutrients in our water systems constitute probably the most costly, widespread, and difficult problem to solve. It is so bad that you can see the problem from space (Figure 14.5). At harmful levels, nitrate, the most oxidized form of nitrogen, is a threat to human and animal health, both directly and indirectly. Nitrates and nitrites in drinking water are regulated by the Safe Drinking Water Act and in wastewater by the Clean Water Act. These government regulations are meant to help protect us against harmful levels of nitrates and nitrites, which can result in methemoglobinemia, a deadly blood disorder. Usually it does not get that bad. More likely is nitrate pollution leading to harmful bacterial and algal growth in wells, lakes, and estuaries (Johnson and Harrison, 2015).

Is the water near you safe to drink? Swim in? Maybe you would like to see how much nitrate is running off a local farm into a nearby stream, river, or lake. Maybe you just want to know if you are putting too much fertilizer on your garden and you could save money by putting on less. You could think about flexing your citizen science muscles and measuring the nitrates yourself. But how?

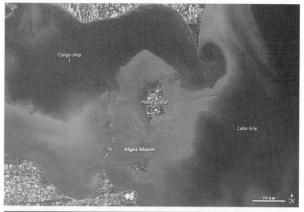

Figure 14.5   On July 28, 2015, *Landsat 8* captured an image of algal blooms in western Lake Erie. NASA Earth Observatory/Landsat. (public domain)

Among the certified methods to detect and quantify nitrates and nitrites is the cadmium reduction method (EPA Method 353.2), which is not what you want, because it requires using and disposing of cadmium, which is a toxic heavy metal. An alternative "green" method for nitrate and nitrite analysis is enzymatic reduction using nitrate reductase (Campbell et al., 2006; Patton and Kryskalla, 2013). This enzymatic nitrate analysis method has been verified to yield results that are equivalent to the accepted EPA-certified methods, but it demands the use of a spectrophotometer. You probably don't have one sitting around the living room. In the old days, such systems were large, bulky, and found only in professional labs. They were also insanely expensive. If you really want to do lab-grade tests of nitrates yourself, what you want is an inexpensive portable photometer to help you detect nitrate pollution.

Luckily for you, the Nitrate Elimination Company (NECi) sponsored the development of an open-source photometer that radically undercut the cost of other methods to detect nitrates with the use of their enzymes (Wittbrodt et al., 2015). Why would the company do this? After all, NECi is primarily a biotech firm that manufactures enzymes. NECi designs enzymes that replace hazardous reagents used for water, soil, food, and fuel analyses. The company is really good at that, but it does not focus on selling tools. To make it easy on itself, NECi harnessed the power of the open-hardware community we talked about in Chapter 13 to make scientific tools. The photometric system the company designed is built on the work that my group had done on open-source colorimeters that could be used to test water quality by measuring chemical oxygen demand (COD) or biochemical oxygen demand (BOD) (Anzalone et al., 2013). This initial device was pretty crude and had to be plugged into a wall outlet (see Figure 14.6).

Figure 14.6 Michigan Tech University student measuring contaminated water for chemical oxygen demand (COD) with an open-source colorimeter. Contaminated water is a major issue in Michigan, but the tests are prohibitively expensive to deploy on a large scale. Here, the open-source device costs about $50 and can be built by students from well-documented plans. It replaces a proprietary device that sells for several thousand dollars. http://www.appropedia.org/Open-source_colorimeter

Subsequent suggestions from the community encouraged us to make a water testing device that could measure turbidity (how cloudy the water is) as well, so we did that next (Wijnen et al., 2014). This tool was then transformed, with the help of NECi, into a spectrophotometer that would work with any Android device to measure nitrates with a degree of accuracy formerly only possible with expensive bench-top equipment. If you are doing just one measurement, you can read it off the screen on your smartphone. If you are doing a major study, data are stored on your Android device and can be downloaded to spreadsheet programs such as Libre Calc or any agricultural software package.

Enzymes "find" their targets at low concentration in complex mixtures, giving you, the citizen scientist, better data with little effort. The reagents are safe for you and the environment. Technically, you could probably drink them (but this is not recommended). Volumes are small. The total assay takes place in 1 milliliter of solution. This creates less waste and fewer collected samples, which are important if you are teaming up with friends to do something like map a river's pollution. Enzyme-based analysis paired with a photometer built around a conventional optical bench and powered by the "brain" of a smartphone running on Android software makes almost anyone capable of generating real, usable nitrate data in the palm of your hand (see Figure 14.7).

NECi now sells test kits based on these enzymes that enable anyone to acquire near-lab-quality results. By agreeing to release the designs of the tool under an open license (Pearce, 2018), the company encouraged new customers, like us citizen scientists, because its enzymes were necessary for the functioning of the device as designed. This invention effectively opened the market of lab-grade nitrate testing to the consumer and citizen scientist level, which had not been possible previously. Although

Figure 14.7 Open-source photometric system for testing nitrates with enzymes. Build instructions and bill of materials available at https://www.appropedia.org/Open -Source_Photometric_System_for_ Enzymatic_Nitrate_Quantification

the company had established customers in the agricultural industry, the device allowed the price point to be pushed down low enough to be used by small family-run farms and even home gardeners. NECi hopes that increased testing of community water can increase public interest in maintaining water quality. Valid data sets are useful for all interested parties, everyone from the government to your lawyer.

References

Anzalone G, Glover A, Pearce J. 2013. Open-source colorimeter. *Sensors* 13(4):5338–5346.

Benchoff B. 2012. Turn your camera phone into a Geiger counter. http://hackaday.com/ 2012/01/15/turn-your-camera-phone-into-a -geiger-counter/

Brasor P. 2012. Public wary of official optimism. *Japan Times*, Sunday, March 11, 2012.

Campbell WH, Song P, Barbier GG. 2006. Nitrate reductase for nitrate analysis in water. *Environmental Chemical Letters* 4:69–73.

Canfield DE, Glazer AN. 2010. The evolution and future of earth's nitrogen cycle. *Science* 330:192–196.

Gundersen A. 2009. Three myths of the Three Mile Island accident, lecture. Fairewinds Energy Education Corp., Burlington, VT.

Gur D, Good WF, Tokuhata GK, et al. 1983. Radiation dose assignment to individuals residing near the Three Mile Island nuclear station. *Proceedings of the Pennsylvania Academy of Science* 57:99–102.

Hatch MC, Beyea J, Nieves JW, Susser M. 1990. Cancer near the Three Mile Island nuclear plant: Radiation emissions. *American Journal Epidemiology* 132(3):397–417.

Hatch MC, Wallenstein S, Beyea J, et al. 1991. Cancer rates after the Three Mile Island nuclear accident and proximity of residence to the plant. *American Journal of Public Health* 81(6):719–724.

Johnson A, Harrison M. 2015. The increasing problem of nutrient runoff on the coast. *American Science* 103:98–101.

Kosmala M, Wiggins A, Swanson A, Simmons B. 2016. Assessing data quality in citizen science. *Frontiers in Ecology and the Environment* 14(10):551–560.

Laureto J, Pearce J. 2016. Nuclear insurance subsidies cost from post-Fukushima accounting based on media sources. *Sustainability* 8(12):1301. https://doi.org/ 10.3390/su8121301

Patton CJ, Kryskalla JR. 2013. Analytical properties of some commercially available nitrate reductase enzymes evaluated as replacements for cadmium in automated, semiautomated, and manual colorimetric methods for determination of nitrate plus nitrite in water. *U.S. Geological Survey Scientific Investigations Report* 36:2013–5033. http://pubs.usgs.gov/sir/2013/5033/

Pearce JM. 2009. Increasing PV velocity by reinvesting the nuclear energy insurance subsidy in large-scale photovoltaic production. In *Proceedings of the 34th IEEE Photovoltaic Specialists Conference (PVSC).* IEEE, Piscataway, NJ, pp. 001338–001343.

Pearce JM. 2012. Limitations of nuclear power as a sustainable energy source. *Sustainability* 4(6):1173–1187.

Pearce JM. 2013. *Open-Source Lab: How to Build Your Own Hardware and Reduce Research Costs.* Elsevier, New York.

Pearce JM. 2018. Sponsored Libre research agreements to create free and open source software and hardware. *Inventions* 3(3):44. https://doi.org/10.3390/inventions3030044

President's Commission on the Accident at Three Mile Island. 1979. *The Need for Change, the Legacy of TMI: Report of the President's Commission on the Accident at Three Mile Island.* President's Commission, Washington, DC.

Talbott EO, Youk AO, McHugh KP, et al. 2000. Mortality among the residents of the Three Mile Island accident area: 1979–1992. *Environmental Health Perspectives* 108: 545–552.

Talbott EO, Zhang A, Youk AO, et al. 2000. Re: "Collision of evidence and assumptions: TMI déjà view." *Environmental Health Perspectives* 108:A547–A549.

Thompson J, Thompson R, Bear D. 1995. *TMI Assessment,* Part 2. http://www.southernstudies.org/images/sitepieces/ThompsonTMIassessment.pdf

U.S. Environmental Protection Agency (EPA). 2015. News release, Newsroom. http://www.epa.gov/adminweb/multimedia/newscontent/2015-2-12-ow/index.html

U.S. Nuclear Regulatory Commission (NRC). 1983. The Price-Anderson Act: The third decade (NUREG-0957). NRC, Washington, DC.

Washington Post. 2011. You can view official EPA radiation readings. *The Washington Post.* http://www.washingtonsblog.com/2011/03/you-can-view-official-epa-radiation-readings.html

Wijnen B, Anzalone GC, Pearce JM. 2014. Open-source mobile water quality testing platform. *Journal of Water, Sanitation and Hygiene for Development* 4(3):532–537.

Wing S, Richardson D. 2000. Collision of evidence and assumptions: TMI déjà view. *Environmental Health Perspectives* 108: A546–A547.

Wing S, Richardson D, Armstrong D, Crawford-Brown D. 1997. A reevaluation of cancer incidence near the Three Mile Island nuclear plant: The collision of evidence and assumptions. *Environmental Health Perspective* 105:52–57.

Wittbrodt BT, Squires DA, Walbeck J, et al. 2015. Open-source photometric system for enzymatic nitrate quantification. *PloS ONE* 10(8):e0134989. https://doi.org/10.1371/journal.pone.0134989

Zelenika-Zovko I, Pearce JM. 2011. Diverting indirect subsidies from the nuclear industry to the photovoltaic industry: Energy and financial returns. *Energy Policy* 39(5): 2626–2632.

CHAPTER 15

Making and Sharing Waste Recycling: Recyclebots

According to the U.S. Environmental Protection Agency (EPA, 2015), the average American produces about 4.5 pounds of waste every day. In nature, there is no such thing as waste; anything generated from one species is simply used as a resource by the next species. Think of mushrooms living on dead trees. There is no reason humans can't learn from nature and follow that design methodology (McDonough and Braungart, 2010). The waste that is generated in your home and business, rather than being a source of loss (e.g., landfill tipping fee), can be treated as a resource. Countless environmental organizations flog "reduce, reuse, and recycle" as a mantra to help the environment by saving money, energy, and natural resources.

This book has already covered a lot about reusing what would be waste as feedstock for our making. We know from Chapter 1 that using an efficient version of Linux, you could resurrect a Microsoft Windows–based computer and then load it up with all kinds of open-source software to make all manner of creations covered in Chapters 2 through 8. Chapter 9 covered some ways to use free patterns and old clothes as starting materials for new, better clothes. Chapter 10 discussed how makers are using waste wood, such as pallets, to make incredible furniture, and Chapter 11 covered how you could use open-source designs to reuse electronics parts. Sometimes waste cannot be simply repurposed and reused, however. In

these instances, the waste needs to be recycled, and Chapter 12, which relied on digital manufacturing techniques, hints at the types of feedstocks that would be the most useful to you as a maker: polymer or composite feedstock and metal wires for 3D printers, sheets of plastic or composite for cutting, and blocks of material for milling. There are many paths to take for this distributed recycling and manufacturing (Dertinger et al., 2020). Let us consider the most mature methods now.

Recycling Waste for 3D-Printer Feedstock

The explosive growth of 3D printing discussed in Chapter 12 brings with it the risk of creating even more un-recycled and wasted plastic. Not every print is perfect. Most people with desktop 3D printers have a box under their desks of failed 3D prints. Could this be a resource as well? The most common types of 3D printers derived from the RepRap use fused-filament fabrication (FFF) and thus demand filament. Fortunately, the economic advantage of distributed manufacturing proven in Chapter 12 increases by an additional order of magnitude with the introduction of recyclebots (Baechler et al., 2013). Recyclebots (www.appropedia.org/Recyclebot) are waste plastic extruders that are used to produce 3D-printer filament from ground-up plastic waste. As you might guess, these tools are pretty good for the environment. Previous research on the life-cycle analysis of distributed 3D printing (Kreiger and Pearce, 2013a, 2013b) already shows that the environment benefits from the reduced transportation of products using traditional manufacturing. The recyclebot process using postconsumer plastics instead of raw materials, however, showed a 90 percent decrease in the embodied energy of the filament from mining, processing of natural resources, and synthesizing compared with traditional manufacturing (Kreiger et al., 2013, 2014). The plastic waste that can be recycled in the United States, shown in Figure 15.1, can all be converted to 3D-printer filament with a recyclebot, but there are many more types of plastic than just the seven identification codes listed in Figure 15.1.

In China, the polymer identification system has seven different classifications of plastic, five different symbols for postconsumer paths, and 140 identification codes (Standardization Administration of the People's Republic of China [SAC], 2008). I was told by the head of the U.S. committee in charge of recycling codes that politics will ensure that the U.S. system is officially trapped in only identifying seven polymers. Luckily, the maker community in the United States is not as challenged. A recent study (Hunt et al., 2015) has already provided a solution for this problem by developing a recycling code model based on the Chinese resin identification codes that is capable of expansion as more complex 3D-printing materials are introduced. If you become the manufacturer, you can tag your own products with the code for

Figure 15.1 Plastic recycling chart. Image courtesy of SouthPack LLC. (Pexel free license) https://www.pexels.com/photo/plastic-recycling-chart-237535/

the plastic you use so that you can recycle it later when you are tired of it or it actually wears out.

With a recyclebot, you can recycle plastic in your own home to save money by offsetting purchased filament, as well as the products you can make with the filament (Zhong and Pearce, 2018). Many versions of recyclebots have been developed by both amateurs and professional makers, such as Matt Rogge from Tech for Trade and even retirees who make them for fun like Hugh Lyman. Major recycling organizations have also made completely open-source versions of recyclebots, including: Plastic Bank, Precious Plastic, and Perpetual Plastic. In addition, there are recyclebots that you can simply purchase, such as the EWE, Extrusionbot, Filastruder, Filafab, Filabot, Filamaker (also has shredder), Noztek, and the Strooder, Felfil (OS). Many of these recyclebots have been used on common 3D-printing failures—so they are comfortable working with polylactic acid (PLA; Cruz Sanchez et al., 2015, 2017), as well as acrylonitrile butadiene styrene (ABS; Mohammed et al., 2017, 2019; Zhong and Pearce, 2018). Recyclebots can also be used for more common plastics, such as high-density polyethylene (HDPE; Baechler et al., 2013), and trickier materials such as flexible elastomers (Woern and Pearce, 2017). Perhaps most excitingly, you can combine wastes in a recyclebot to make new materials. For example, you can take waste saw dust and add it to PLA to make waste-wood composites (Pringle et al., 2017) that you can use for 3D printing. This early work, however, hardly begins to scratch the surface of the potential to use distributed methods to recycle a much longer list of polymers, as well as composites made up of multiple distributed waste streams. Recently, to carry on the open-source design mentality (Oberloier and Pearce, 2018) in hardware espoused by the RepRap community, my research group and I made a RepRapable

recyclebot. This type of recyclebot, shown in Figure 15.2, can be largely manufactured in your living room with any of the types of RepRaps discussed in Chapter 12. It works for a wide range of common 3D-printing plastics such as PLA and ABS.

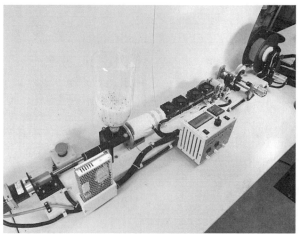

Figure 15.2 RepRapable recyclebot, a waste plastic extruder that makes 3D-printer filament, which can be used to print most of its own parts. (GNU GPL) https://www.appropedia.org/File:Recyclebotrep.png

Recyclebots can even be mounted directly on 3D printers. This is what re:3D, an open-source 3D-printer company made up of a bunch of NASA engineers, has done. My lab, in conjunction with re:3D, has demonstrated the ability to directly print large products with what re:3D calls the *GigabotX* (shown in Figure 15.3) from waste from PLA, ABS, polypropylene (PP), and polyethylene terephthalate (PET [what makes up water bottles]; Woern et al., 2018b) and advanced stronger materials such as polycarbonate (PC; Reich et al., 2019). Being able to print in PC is important because of its higher melting temperature and strength. This allows you to print molds for lower-melting-temperature plastics. Thus, if there is something you want a lot of—or if you want a really large solid object—it is much easier to print the mold

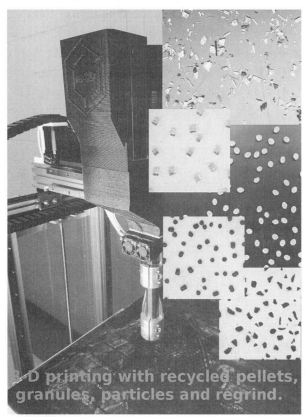

3-D printing with recycled pellets, granules, particles and regrind.

Figure 15.3   The GigabotX, an open-source plastic pellet 3D printer. (GNU GPL) https://www.appropedia.org/File:Gigarecycle.png

(just the outside) and then use a recyclebot or injection molder to fill it. Ultimaker's open-source Cura slicing software already has a default mold-making slice, so if there is a 3D object for which you want to make a mold, you don't even need to specifically computer-aided design (CAD) it because you can start with an STL file.

To assist with this type of recycling, my student and I also developed the sophisticated pelletizer seen in Figure 15.4 (Woern and Pearce, 2018). This pelletizer can be made with a drill and some 3D-printed parts to enable fractional mixing of several material streams (up to four with the current design). This gives you the power to make sophisticated, controlled composites with different plastics or entirely different types of waste, such as paper, cardboard, sawdust, powered metal, glass, and so on.

Printing with waste plastic even improves the economics further for 3D printing your own products. As we discussed in Chapter 12, you can normally save 90 to 99 percent of the total cost of an item when 3D printing a product with commercial filament than when buying it

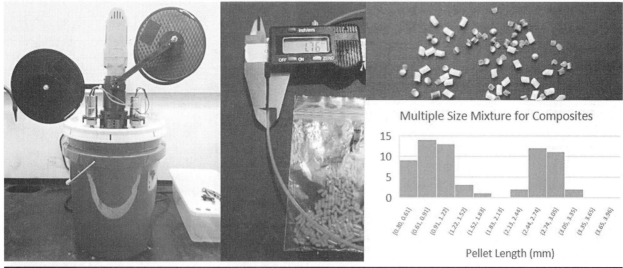

Figure 15.4   An open-source pelletizer that can be used to make sophisticated feedstock mixtures for composites in 3D printing, sheet forming, or block molding. (GNU GPL) https://www.appropedia.org/File:OSpellets.JPG

from a store or online. Commercial filament has a huge markup and costs anywhere from about $15 to $80 per kilogram. You can make filament with a recyclebot from pellets for less than $5 per kilogram. Better yet, if you shred your own waste, the cost is only the electricity to run the machines, which drops the cost below 10 cents per kilogram. If you use insulation on the hot end of your recyclebot, this value can come down to only 2 or 3 cents per kilogram, depending on how much you pay for electricity. If you happen to have built a solar-powered recyclebot (Zhong et al., 2017) or 3D printer (King et al., 2014; Gwamuri et al., 2016), the marginal cost of the whole process is essentially free to you. Critics will point out that I didn't factor in a time cost. Of course, it will take you some time to make the products you want, but normal people don't charge for their own time to do things for themselves. When was the last time you calculated the opportunity cost to butter your toast in the morning? Or if you go to McDonald's, do you calculate the opportunity cost to drive yourself? Even Warren Buffett (net worth $74 billion) drives himself to McDonald's for breakfast every morning on his way to work (Elkins, 2017). Not that this is his best idea ever—so don't copy that one!

When you make things yourself, the savings get simply silly. Consider the whole process shown in Figure 15.5 for making a camera lens cover. It is made out of ABS, the same plastic as Legos. There is a lot of ABS waste in anything computer related. The source of the ABS used in this case was from e-waste from my university's information technology (IT) department dumpster. It was ground up in an open-source granulator (Ravindran et al., 2019), made into filament by a vertical recyclebot, and then printed into a lens cover. The total cost in electricity was only 3 cents! An equivalent camera lens hood costs $9.99 on Amazon. Thus, you could make 333 camera lens hoods for the same economic cost as purchasing a single one (Zhong and Pearce, 2018). Lastly, I should point out that you are not limited to making only small things. It all depends on the size of the printer you can get access to. Today, many libraries, community centers, makerspaces, fablabs, and machine shops have 3D printers on which you can start to make large products. For example, the kayak paddles that you can custom make for yourself (Figure 15.6) can be made for a small fraction of the cost of high-end kayak paddles (Byard et al., 2019).

Figure 15.5 The whole process of waste plastic conversion to a product through a vertical recyclebot. (GNU GPL) https://www.appropedia.org/File:Recycledabs.jpg

Figure 15.6 3D-printed kayak paddles made from waste ABS. (GNU GPL) https://www.appropedia.org/File:Kayak3dp.jpg

Sheets, Blocks, and Bulk Extrusions

You can use recyclebots as compounders and the pelletizer to produce plastic for making sheets, blocks, or bulk extrusions. Although you might want to own a 3D printer yourself to manufacture products for yourself (and the average person may even consider a recyclebot), for larger, more expensive tools such as a GigabotX, hot presses, or large extruders, it may make more sense to use them in a shared facility.

For example, Waste for Life (www.wasteforlife.org) is a loosely joined network of makers who work together to develop poverty-reducing solutions to specific ecological problems. One of its projects involved collaboration with researchers and community members in Canada, Argentina, the United States, Australia, and Lesotho to build the open-source Kingston Hot Press to provide the means of production to smaller cooperatives in communities anywhere in the world. A hot press allows the user to produce a value-added composite tile out of waste plastic and fiber, most commonly cardboard and paper. You can use a hot press to start to recycle waste plastic for larger, bulkier projects for which you need a good source of sheeting. The sheets can be made into millions of products by either heat forming them by hand or by using something like an open-source CNC cutter to cut them into 2D shapes for assembly.

Similarly, Precious Plastic (preciousplastic.com) is a global community of hundreds of people working toward a solution to plastic pollution. All of its recycling solutions are free and open source as well. Precious Plastic's systems are designed to be modular, and you can construct them yourself if you have access to a decent machine shop. If not, the group runs an online bazaar where people who have access to welding and metal-cutting tools will make the parts or the whole machines and sell them for reasonable prices. Precious Plastic currently has shredders, an extruder (recyclebot), an injection molder, and a hot compression machine. By making large metal molds and attaching them directly to the extrusion machine, the group can make large recycled plastic blocks or even plastic timber (just like the really expensive stuff some people use for decking).

The tools for plastic recycling at the distributed home user scale are relatively mature and well-documented, as shown earlier. For other materials, such as glass and metal, the communities are smaller and the technologies less advanced. Sand is relatively inexpensive, so recycling glass in an economical way, even at the industrial scale, is challenging. There are some projects that you may be interested in if you have some top-tier maker skills, such as the Kube OpenLathe, which is an open-source glass-blowing lathe that was funded on Kickstarter. Open Source Ecology (www.opensourceecology.org), a nonprofit group I will talk a lot more about in Chapter 16, is also working on an open-source induction furnace that will bring metal recycling much closer to home in the future.

References

Baechler C, DeVuono M, Pearce JM. 2013. Distributed recycling of waste polymer into RepRap feedstock. *Rapid Prototyping Journal* 19(2):118–125.

Byard DJ, Woern AL, Oakley RB, et al. 2019. Green fablab applications of large-area waste polymer–based additive manufacturing. *Additive Manufacturing* 27:515–525.

Cruz Sanchez F, Boudaoud H, Hoppe S, Camargo M. 2017. Polymer recycling in an open-source additive manufacturing context: Mechanical issues. *Additive Manufacturing* 17:87–105.

Cruz Sanchez F, Lanza S, Boudaoud H, et al. 2015. Polymer recycling and additive manufacturing in an open source context: Optimization of processes and methods. In *Proceedings of the 2015 Annual International Solid Freeform Fabrication Symposium: An Additive Manufacturing Conference*, pp. 10–12.

Dertinger SC, Gallup N, Tanikella NG, et al. 2020. Technical pathways for distributed recycling of polymer composites for distributed manufacturing: windshield wiper blades. *Resources, Conservation and Recycling* 157:104810. https://doi.org/10.1016/j.resconrec.2020.104810

Elkins K. 2017. Warren Buffett eats the same thing for breakfast every day—and it never costs more than $3.17. CNBC. https://www.cnbc.com/2017/01/30/warren-buffetts-breakfast-never-costs-more-than-317.html

Environmental Protection Agency (EPA). 2015. Advancing sustainable materials management: 2014 fact sheet. https://www.epa.gov/sites/production/files/2018-07/documents/2015_smm_msw_factsheet_07242018_fnl_508_002.pdf

Gwamuri J, Franco D, Khan K, et al. 2016. High-efficiency solar-powered 3D printers for sustainable development. *Machines* 4(1):3. https://doi.org/10.3390/machines4010003

Hunt EJ, Zhang C, Anzalone N, Pearce JM. 2015. Polymer recycling codes for distributed manufacturing with 3D printers. *Resources, Conservation and Recycling* 97:24–30.

King DL, Babasola A, Rozario J, Pearce JM. 2014. Mobile open-source solar-powered 3D printers for distributed manufacturing in off-grid communities. *Challenges in Sustainability* 2(1), 18–27.

Kreiger M, Anzalone GC, Mulder ML, et al. 2013. Distributed recycling of post-consumer plastic waste in rural areas. *MRS Online Proceedings Library Archive* 1492:91–96.

Kreiger M, Pearce JM. 2013a. Environmental life cycle analysis of distributed three-dimensional printing and conventional manufacturing of polymer products. *ACS Sustainable Chemical Engineering* 1: 1511–1519.

Kreiger MA, Pearce JM. 2013b. Environmental impacts of distributed manufacturing from 3D printing of polymer components and products. *MRS Online Proceedings Library Archive* 1492:mrsf12-1492-g01-02.

Kreiger MA, Mulder ML, Glover AG, Pearce JM. 2014. Life cycle analysis of distributed recycling of post-consumer high density polyethylene for 3D printing filament. *Journal of Cleaner Production* 70:90–96.

McDonough W, Braungart M. 2010. *Cradle to Cradle: Remaking the Way We Make Things*. North Point Press, New York.

Mohammed MI, Mohan M, Das A, et al. 2017. A low carbon footprint approach to the reconstitution of plastics into 3D-printer filament for enhanced waste reduction. *KnE Engineering* 2:234–241.

Mohammed M, Wilson D, Gomez-Kervin E, et al. 2019. Investigation of closed loop manufacturing with acrylonitrile butadiene styrene (ABS) over multiple generations using additive manufacturing. *ACS Sustainable Chemistry and Engineering* 7(16), 13955–13969.

Oberloier S, Pearce JM. 2018. General design procedure for free and open-source hardware for scientific equipment. *Designs* 2(1):2.

Pringle AM, Rudnicki M, Pearce JM. 2018. Wood furniture waste-based recycled 3D printing filament. *Forest Products Journal* 68(1):86–95. https://doi.org/10.13073/FPJ-D-17-00042

Ravindran A, Scsavnicki S, Nelson W, et al. 2019. Open source waste plastic granulator. *Technologies* 7(4):74. https://doi.org/10.3390/technologies7040074

Reich MJ, Woern AL, Tanikella NG, Pearce JM. 2019. Mechanical properties and applications of recycled polycarbonate particle material extrusion-based additive manufacturing. *Materials* 12(10):1642. https://doi.org/10.3390/ma12101642

Standardization Administration of the People's Republic of China (SAC). 2008. 2008 *Marking of Plastics Products* (GB16288). Chinese Standard Publishing House, Beijing.

Tian X, Liu T, Wang Q, et al. 2017. Recycling and remanufacturing of 3D printed continuous carbon fiber reinforced PLA composites. *Journal of Cleaner Production* 142:1609–1618. https://doi.org/doi:10.1016/j.jclepro.2016.11.139

Woern AL, Byard DJ, Oakley RB, et al. 2018. Fused particle fabrication 3D printing: Recycled materials' optimization and mechanical properties. *Materials* 11:1413. https://doi.org/10.3390/ma11081413

Woern AL, McCaslin JR, Pringle AM, Pearce JM. 2018. RepRapable recyclebot: Open source 3D printable extruder for converting plastic to 3D printing filament. *HardwareX* 4:e00026. https://doi.org/10.1016/j.ohx.2018.e00026

Woern AL, Pearce JM. 2017. Distributed manufacturing of flexible products: Technical feasibility and economic viability. *Technologies* 5(4):71. https://doi.org/10.3390/technologies5040071

Woern AL, Pearce JM. 2018. 3D printable polymer pelletizer chopper for fused granular fabrication-based additive manufacturing. *Inventions* 3(4):78. https://doi.org/10.3390/inventions3040078

Zhong S, Pearce JM. 2018. Tightening the loop on the circular economy: Coupled distributed recycling and manufacturing with recyclebot and RepRap 3D printing. *Resources, Conservation and Recycling* 128:48–58.

Zhong S, Rakhe P, Pearce JM. 2017. Energy payback time of a solar photovoltaic powered waste plastic recyclebot system. *Recycling* 2(2):10. https://doi.org/10.3390/recycling2020010

CHAPTER 16

Making and Sharing Big Free Stuff

Introduction to Free Big Stuff

We know that free and open-source sharing does not have to stop with immaterial items of value such as code and software; it also can be used to improve innovation and save money on hardware. Many open-hardware projects are small. They could fit in a "bread box"—or at least on a desk. Yet there is no reason that we cannot make large, free hardware as well. In this chapter, we will look at hardware designs for cars, ways to cut the cost of electricity, the Open Building Institute (which is developing a range of house designs),

and Open Source Ecology (which is trying to establish the open-source blueprints for the machines that drive civilization itself).

Open-Source Cars

Although the younger generation is often opting to avoid car ownership in favor of ride-sharing services, many people love cars. If you really love cars and would like to help design and build your own, then the Rally Fighter is the project for you! The Rally Fighter (Figure 16.1) was developed by Local Motors (localmotors.com) after only

Figure 16.1 Rally Fighter. Image by C. Blizzard. (CC BY-SA) https://www.flickr.com/photos/christopherblizzard/5741142315/in/photostream/

18 months of development. This constitutes a record *time to market* compared with automobile industry standards, which are about four times longer. Local Motors was able to bring their car to market so rapidly by applying innovative technologies and open-source crowdsourcing techniques. The Rally Fighter is notably the first production vehicle designed via crowdsourcing (i.e., asking for help from the global community on the internet). The winning design, as a result of a vote from the community, was submitted by Sangho Kim.

In addition to its open-source model of development, Local Motors also has an innovative method to sell the vehicles. The company sells cars as kits that you can then help build! You go to one of Local Motor's microfactories and assemble your car with help from the company's team. This also allows the Rally Fighter to be titled a *kit car* in the United States, so it can be sold without the normal regulations although to be clear, it is street legal in the entire United States. Local Motors is a small automobile company, but it has already made great headway on its next project—Olli, the world's first cocreated self-driving electric and cognitive shuttle.

Open source is appealing to large automobile companies as well. In 2014, the CEO of Tesla, Elon Musk, announced a radical new patent pledge, saying, "Tesla irrevocably pledges that it will not initiate a lawsuit against any party for infringing a Tesla patent through activity relating to electric vehicles or related equipment for so long as such party is acting in good faith" (Tesla, 2014). Musk explained that the reason the company did this was that "[t]echnology leadership is not defined by patents, which history has repeatedly shown to be small protection indeed against a determined competitor, but rather by the ability of a company to attract and motivate the world's most talented engineers. We believe that applying the open source philosophy to our patents will strengthen rather than diminish Tesla's position in this regard" (Musk, 2014). This position has been clarified: Tesla really allows others to use its electric vehicle patents without a formal license (Masnick, 2015). Tesla is not alone—even Toyota has joined the Linux Foundation, a major open-source organization (Vaughan-Nichols, 2011).

OpenEVSE (www.openevse.com) is another open-source company focused on automobiles. This company provides open-source solutions for electric vehicle charging.

If you have not caught the e-car bug yet and you have a "normal" car, you can still add a little open-source magic. For example, OpenXC (openxcplatform.com) is an open-source hardware and software platform that provides a data-focused application programming interface (API) for your car. By installing the OpenXC hardware module, your vehicle data become accessible to open-source Android or other desktop applications using the OpenXC library. OpenXC supports the U.S. federally mandated "on-board diagnostics" data on almost all cars and over an even wider range of data sets on most modern Ford vehicles. This means you can monitor many of the sensors on your car or truck so you can enable new and innovative vehicle-centric applications, learn how to improve your vehicle's performance, or teach your kids about cars. It even allows you to perform your own diagnostics—no more taking it to the mechanic because the mysterious check engine light is on.

Free Electricity

There is free hardware for several methods of generating electricity, including wind power (OpenSourceLowTech, 2019), but for many sources of renewable power, you have to have a good site, located either on a river or stream or in a nice windy location. Not everyone has this, but most everyone gets a little sunshine.

This section therefore concentrates on solar energy and solar photovoltaic (PV) technology in particular. PV devices convert sunlight directly into electricity. Recently, the costs have dropped so radically that PV devices are now the least expensive way of generating electricity (International Renewable Energy Agency [IRENA], 2019). Now—not surprisingly—PV devices have been the fastest-growing energy source for some time (Vaughan, 2017). Although most PV devices are on the grid and part of major solar farms owned by electric utilities, PV systems can be designed for very small applications to provide distributed power generation. For a complete guide to such small systems, see the book *To Catch the Sun* (Grafman and Pearce, 2020). PV systems have no moving parts, so there is nothing to break.

PV systems are also modular. Need more power? Just add another panel. Sunlight is free, so once you purchase the system, you get electricity for 20 years or more. No noise or pollution is created from operating PV systems, so they are really good for the environment. Open-source approaches to PV technology could help scientists to accelerate its development (Buitenhuis and Pearce, 2012), but there are a lot of open-source aspects of PV systems that anyone can use now. For example, many open-source designs are being used to charge cell phones, as we saw in Chapter 2 (see Figure 2.3). There are also scores of open-source electronics designs using PV devices to power everything from toys (ASCAS, 2019) to Chapter 15's open-source recyclebots (Zhong et al., 2017) and Chapter 12's 3D printers. An example of the

(a)   (b)

Figure 16.2 (a) Photograph of the MOST-delta 3D printer and PV stand-alone power system assembly in transit in a duffel bag. For transportation, the modules are mounted on all three sides of the delta. (b) The same system deployed on a picnic bench. (CC BY-SA) https://www.appropedia.org/High-Efficiency_Solar-Powered_3-D_Printers_for_Sustainable_Development

latter is the solar-powered delta-style RepRap that fits in a duffle bag (Figure 16.2) and helps you manufacture the things you need pretty much anywhere (Gwamuri et al., 2016).

If you want to power something big, like your house, you need to figure out how many solar panels you need. There are several ways to do free, detailed engineering studies, including the Open Source Solar Advisory Model (SAM) from the U.S. National Renewable Energy Lab (NREL, 2019) (github.com/NREL/SAM). If you just want to do a quick calculation for your home, you are free to use PVWatts (pvwatts. nrel.gov). It is not as powerful as SAM (nor open source, although it uses the same basic equations), but it is good enough to help you calculate approximately how much money you will save.

PV systems generally consist of a PV panel, an inverter (which converts battery-like direct-current [DC] power to wall-outlet-like alternating-current [AC] power), and racking to hold the panels up. It is possible to be off-grid in an isolated system that uses battery backup. Most PV systems today, however, are plugged into the grid so that if you are making solar energy you don't need, you can share it with your neighbors.

In the not-so-distant past, PV technology was expensive. Not anymore! Today, the cost of PV modules has come down so far that the most expensive part of the system is the rack. For a super-cheap rack for a small, temporary system, you can just prop a panel up against something like a stone or the wall of your house. If you are interested in something a little more permanent and can't quite afford an amazing solar-powered Tesla roof, open-source designs are available for other building-integrated PV (BIPV) systems (Pearce, Meldrum, and Osborne, 2017). BIPV systems eliminate the need for roofing—the solar panels become the roof. Depending on how you look at it, either you cut your roofing cost or the roof cuts your PV system costs. Ground-mounted systems can be built from treated wood or simple pipes rather than proprietary metal racks for pennies on the dollar.

Open Building Institute

What if you want more than your roof to be open source, sustainable, and cost-effective? You could have a completely open-source house, developed by the Open Building Institute (OBI; www.openbuildinginstitute.org). The basic functionality and details of OBI open-source microhouses are shown in Figure 16.3. There are even open-source plans to monitor all the electric loads and PV generation in the house (Oberloier and Pearce, 2018).

These houses are not only solar-powered, they also collect their own waste water and can be built by you and your friends in multiday microhouse extreme manufacturing workshops. They are full featured, right sized, and classy, and even have a built-in aquaponic greenhouse that can grow food for your family (Figure 16.4). The greenhouse uses passive solar heating for two in-ground fish ponds, a chicken coop connected to an outside run, a rabbit pen, 44 six-foot-tall aquaponic towers, 216 feet of aquaponic troughs, two compost beds, and a potting and seedling area. The greenhouse is designed to produce all the vegetables, mushrooms, and fish that a family might need.

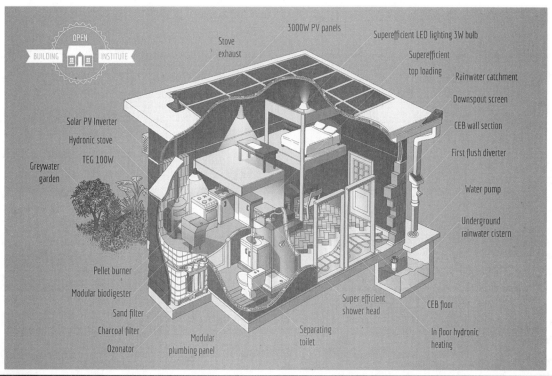

Figure 16.3   Open-source microhouse infographic by Jean-Baptiste Vervaeck, Open Building Institute. (CC BY-SA) https://www.openbuildinginstitute.org/how-it-works/

Figure 16.4   Open-source aquaponic greenhouse, Open Building Institute. (CC BY-SA) https://www.openbuilding institute.org/buildings/

Open Source Ecology

Open Source Ecology (OSE; www.opensource ecology.org) actually developed many of the tools OBI needs to fabricate houses. The goal of OSE is to provide open-source blueprints for civilization itself, wherever you happen to live. OSE is doing this by providing the open-source plans for the Global Village Construction Set (Figure 16.5), which consists of 50 of what OSE thinks are the most important devices for modern society. Each open-source device is meant to be modular, able to be built by you, and low cost. Overall, the idea is to have a high-performance platform that allows for the easy fabrication of the 50 industrial machines that it takes to build a small, sustainable civilization with modern comforts. Similar to the lessons we have seen elsewhere in this book, open-source solutions cost a fraction of commercial solutions.

There has been steady development of these 50 tools with help from specific open-source communities, including the laser cutter and 3D printers (see Chapter 12). OSE, however, has taken the lead on many of the tools, such as a tractor, backhoe, trencher, ironworker, rototiller, soil pulverizer, and compressed-earth block press, already prototyped and, in several cases (e.g., the tractor in Figure 16.6), replicated all over the world. The 27-horsepower open-source LifeTrac tractor looks a lot like it was built from adult-sized Legos or Tinkertoys. This is exactly the idea! By using a modular design, components can be reused or swapped. Thus, for example, the *power cube*, which is a modular self-contained universal power unit consisting of an engine coupled with a hydraulic pump, provides power from hydraulic fluids at high pressure through quick-connect hoses. It can be used to power the tractor, but also can be taken off to power other OSE machines.

Figure 16.5   The 50 open-source devices in the Global Village Construction Set developed by Open Source Ecology. (CC BY-SA) https://www.opensourceecology.org/gvcs/

Figure 16.6 Open-source tractor—the LifeTrac 6—developed by Open Source Ecology. (CC BY-SA) https://opensourceecology.dozuki.com/c/LifeTrac

References

ASCAS. 2019. DIY mini solar car (14kph toy)! Instructables. https://www.instructables.com/id/DIY-Mini-Solar-Car-14kph-Ultrafast-Toy/

Buitenhuis AJ, Pearce JM. 2012. Open-source development of solar photovoltaic technology. *Energy for Sustainable Development* 16: 379–388. http://dx.doi.org/10.1016/j.esd.2012.06.006

Grafman L, Pearce JM. 2020. *To Catch the Sun.* https://www.appropedia.org/To_Catch_the_Sun

Gwamuri J, Franco D, Khan KY, et al. 2016. High-efficiency solar-powered 3D printers for sustainable development. *Machines* 4(1):3. http://dx.doi.org/10.3390/machines4010003

International Renewable Energy Agency (IRENA). 2019. Renewables now the lowest-cost power source in most of the world. Institute for Energy Economics and Financial Analysis. http://ieefa.org/irena-renewables-now-the-lowest-cost-power-source-in-most-of-the-world/

Masnick M. 2015. Elon Musk clarifies that Tesla's patents really are free; investor absolutely freaks out. *Techdirt.* https://www.techdirt.com/articles/20150217/06182930052/elon-musk-clarifies-that-teslas-patents-really-are-free-investor-absolutely-freaks-out.shtml

Musk E. 2014. All our patents belong to you. https://www.tesla.com/blog/all-our-patent-are-belong-you

SAM Open Source. 2019. System Advisor Model (SAM). https://sam.nrel.gov/about-sam/sam-open-source.html

Oberloier S, Pearce JM. 2018. Open source low-cost power monitoring system. *HardwareX* 4:e00044. https://doi.org/10.1016/j.ohx.2018.e00044

OpenSourceLowTech. 2019. Wind turbine tutorial. OpenSourceLowTech.org. http://opensourcelowtech.org/wind_turbine.html

Pearce JM, Meldrum J, Osborne N. 2017. Design of post-consumer modification of standard solar modules to form large-area building integrated photovoltaic roof slates. *Designs* 1(2):9. http://dx.doi.org/10.3390/designs1020009

Tesla. 2014. Patent ledge: privacy and legal. Tesla. https://www.tesla.com/about/legal

Vaughan A. 2017. Time to shine: Solar power is fastest-growing source of new energy. *The Guardian.* https://www.theguardian.com/environment/2017/oct/04/solar-power-renewables-international-energy-agency

Vaughan-Nichols SJ. 2011. The open-source car: Toyota is joining the Linux Foundation. *ZDNet.* https://www.zdnet.com/article/the-open-source-car/

Zhong S, Rakhe P, Pearce JM. 2017. Energy payback time of a solar photovoltaic powered waste plastic recyclebot system. *Recycling* 2(2):10. https://doi.org/10.3390/recycling2020010

CHAPTER 17

How to Make
a Million Dollars of Value

As we discussed in Chapter 12, digital manufacturing technologies, such as 3D printing, have matured a lot (Gershenfeld, 2005)—to the point where you can use them in your living room or at least your garage, no matter where you live (Pearce et al., 2010). There has been an exponential rise in designs for hardware released under open-source Creative Commons licenses or placed in the public domain (Wittbrodt et al., 2013). This enables you to take advantage of a new paradigm: distributed manufacturing of all kinds of products released under free licenses that we call *free and open-source hardware* (FOSH). The availability of these designs has a large value to those with access to digital manufacturing methods. But how much value? In this chapter, we will look at the simplest ways of calculating this value and show you how to spend some of your time to create a million dollars of value for the world!

A Wee Bit of Math

This chapter is based on a study I used to justify investment in FOSH for scientists (Pearce, 2015), but the math can be useful to you for calculating how much value you create for the world with any type of open-source product. I will discuss two ways to quantify the value of FOSH designs: (1) *downloaded substitution valuation* and (2) *avoided reproduction valuation*. The first

is the easiest to do, so I will focus on this option because it is (a) relatively straightforward to understand, (b) based on reliable, freely available data, and (c) minimizes assumptions. All the straightforward ways to calculate value can be expanded further with additional (harder to nail down) benefits related to market expansion, scientific innovation, acceleration (e.g., all the scientific devices discussed in Chapter 13), educational enhancement (e.g., can the protractor you designed be used by students in school to understand geometry), and health benefits that increase longevity and productivity (e.g., your design for a new sit-up machine could help others increase their core strength).

Downloaded Substitution Valuation

The downloaded substitution valuation uses the number of times that an open-source design is downloaded to quantify the value of the design to the world. The basic idea is that the value of the hardware design is equal to the number of times it was downloaded times the difference in the cost of the open hardware and the commercial hardware. Thus, the downloaded valuation for substitution savings V_D at time t can be written as

$$V_D(t) = S \times P \times N_D(t) \qquad (17.1)$$

where the savings S from FOSH digital manufacturing is determined simply by subtracting the cost to make something from what it would cost to buy. Here, S has been shown to be substantial in the developing world for appropriate technology (Pearce et al., 2010), as well as for consumer goods (Petersen and Pearce, 2017) and even toys (Petersen et al., 2017). Moreover, S is maximized for custom low-volume products such as scientific or medical equipment, where the open-source cost is generally only 1 to 10 percent of the cost to buy a commercial product, as we learned in Chapter 13 (Pearce, 2014a). In the equation, $N_D(t)$ is the number of times the digital design has been downloaded at time t. Many websites keep track of this number for you.

You will notice that there is one extra term, P, which is the percent of downloads resulting in a product. We do not really know exactly what value to put in for P because some people could download the designs and not use them or, perhaps more likely, a person could download a design and make a bunch of copies to make a set, exchange via email, put on a memory stick, or post on peer-to-peer (P2P) sites that are not recorded. In scientific studies, we use $P = 1$. Thus, it does not impact the equation at all (anything multiplied by one is that number). We can say that $P = 1$ based on my informal discussions with hundreds of RepRap owners in which they claim that the vast majority of designs downloaded were printed. You can think of it like an mp3 file that you might download from the web. You probably listened to the file at least once. It is the same for files to laser cut or to 3D print. There is no value in them to the user unless the user uses them to make the thing the file describes. This assumption has some error associated with it, but it is a conservative assumption (meaning that your estimates of value will be underestimates). Also, S can change with time, and the number of total downloads tends to increase as time marches on. Money has different value based on what time you are talking about, so if you really want to get into the nitty-gritty details, see the paper (Pearce, 2015). To get a good estimate of the value of your design, just take the difference in cost to make it versus to buy it and multiply by the number of downloads.

Avoided Reproduction Valuation

What if the thing you want to design and share is simply not available anywhere on the market? How do you calculate the value then? We can do this using the avoided cost of reproduction value V_R for a single buyer. The idea is that you could determine the value to make something new that is not really on the market yet based on how much it would cost to hire someone to make it. Thus V_R is given by multiplying h (the number of design hours needed to replicate the product) and w (the hourly wage of the workers needed to produce the product). The more complicated the design, the harder (more skill) and longer it takes you to make it, the greater the value. Mathematically, this is shown as

$$V_R = h \times w \qquad (17.2)$$

This method of capturing value can also be extrapolated to everyone (e.g., individuals who would hire firms or freelance designers to complete the design) to obtain the total value to society. Again, you can take variations in wages and discount-rate variables into account, but the reader is referred to my paper on the subject (Pearce, 2015) if you really want to get into the details.

Case Studies

With a few real-world examples, everyone can see how they can make a million dollars of value. The first is a case study to determine the value of an open-source syringe pump design (Wijnen et al., 2014). In this case, the syringe pump represents a valuable tool that may be funded by the government for the acceleration of scientific innovation, but that also has applications in education and medicine. The other case studies include sporting goods: a compound bow balancer, a dead-lift jack, and a whistle.

Open-Source Syringe Pump

Figure 17.1 shows an example of a pump you can make using an open-source syringe pump (OSSP) library (Wijnen et al., 2014). The library was designed using open-source and freely available OpenSCAD (openscad.org). Most of the pump parts can be fabricated with an open-source RepRap 3D printer (reprap.org), whereas other necessary parts are readily available, such as a stepper motor and steel rods. The design, bill of materials, and assembly instructions are globally available to anyone wishing to use them (www.appropedia.org/Open-source_syringe_pump). You can use your cell phone to drive the device with wireless control because it runs on a quasi-open-source Raspberry Pi minicomputer (www.raspberrypi.org).

The original study on the syringe pump found that it is as good as (or better than) commercial syringe pumps. The low-cost open-source variety of syringe pumps, however, are completely customizable, allowing both the volume and the motor to scale for specific applications, such as any research activity, including the carefully controlled dosing of reagents and pharmaceuticals and the delivery of viscous 3D-printer media. So how much is it worth?

Method 1: Downloaded Substitution Value

The designs for the open-source pump were released in September of 2014, and in one month, the designs had been downloaded from two digital repositories a total of N_D = 424 times (114 on Thingiverse and 310 on Youmagine). The cost to purchase a traditionally manufactured syringe pump C_p ranges from $260 to $1,509 for a single pump and $1,800 to $2,606 for a dual pump (Wijnen et al., 2014). The cost of the materials for a single open-source pump is $97, and for the double, it is $154. The time to assemble either the single or double pump is less than an hour and can be

Figure 17.1 Michigan Tech student showing off her variant of the open-source syringe pump. (CC BY-SA) https://www.appropedia.org/Open-source_syringe_pump

accomplished by a nonexpert. Although the time to print the components is less than four hours on a conventional RepRap, workers can do other tasks while printing. The assembler hourly rate is assumed to be $10 per hour because no special skills are needed. As discussed earlier, it was assumed that $P = 1$.

This provides a savings for substituting the open-source syringe pump for a commercial one of between $153 and $1,402 for a single pump and between $1,636 and $2,442 for a double pump. Thus, following Equation (17.1), the value V_D of the pump library after one month ranged from more than $64,000 to more than $1,000,000 for the global community. A million dollars of value in a month! Not bad. Performing even a simple linear extrapolation for a single year provides a total value of between $778,000 and $12.4 million. Obviously, this calculation can be expanded for more years, but then the discount rate must be taken into account, which can vary depending on the organization or individual performing the analysis.

There is a huge uncertainty in the value saved based on what the syringe pump is replacing—is it a high-end or low-end pump? The low-end value is based on a simple infusion pump with considerably less functionality than the open-source syringe pump. Although it is possible that some of the downloaders only needed a simple infusion pump, the majority would be likely to be replacing more sophisticated devices. In addition, the open-source library allows for more programmable control than any of the other pumps on the market; thus, it appears reasonable to estimate the value using a midrange pump such as the GenieTouch Syringe Pump for $675, which provides a savings of $568 per pump. This provides a value of over $5 million today (based on around 10,000 downloads) for the global commons. In addition, using the OSSP library, it is possible to make even more sophisticated and valuable equipment. For example, for $308

in parts, you can construct a four-syringe pump using the FOSH design library, but a Cole-Parmer Continuous Flow Syringe Pump with four syringes costs $3,947, which is an S value of more than $3,600 for a single download!

Method 2: Avoided Reproduction Valuation

The mechanical designs for the open-source syringe pump were completed by experienced engineers in less than 6 worker-hours. To print and revise the five 3D-printed components took 3 hours, assembly took less than 1 hour, and software development and Pi wiring took less than 16 hours. The total design and prototyping time was less than 26 hours. This schedule was possible because the designers were experienced with similar designs and had access to all the components. According to Glassdoor, the median annual salary for a CAD engineer and a software engineer is approximately $89,000, and ignoring overhead and benefits costs to be conservative for a 50-week year, working 40 hours per week, the w value is $44.50. Thus, V_R from Equation (17.2) is $1,157. With the cost of the parts for a single open-source syringe pump being $97, the total cost for the first pump for a firm designing it is $1,254, which is more expensive than the lowest-cost syringe pumps, but still would provide savings for the high-range products. Thus, it is likely that some companies have designed in-house pumps before, and there is some evidence of this (i.e., syringe pump singular designs at Massachusetts Institute of Technology, Hackaday, and Openpump.org). These other designs were not global pump libraries, although they would provide solutions for a subset of syringe pump users. Now, using $N_D = 424$ as a proxy for people who may be willing to either complete the design themselves or hire a freelancer, after the first month, the value is more than $490,000, and using the same extrapolation to the first year, it would be more than $5.8 million.

It has been five years, and it is instructive to look at the values of wealth actually generated by the open-source syringe pump, which is easily many millions of dollars. Because the open-source syringe pump represents a savings of between 59 and 93 percent, if it gains significant market penetration, it can be expected to increase the market for syringe pumps. Because the pump meets the standards for research and has already been vetted, it seems reasonable that it would be most likely to be adopted by university labs first. This appears to be what has happened based on the number of derivatives of the open hardware found throughout university labs. These derivations can become quite significant, such as with the recently open-source Y-struder shown in Figure 17.2 (Klar et al., 2019). Thus, there is potential additional value that the FOSH design provides for both scientific research and education.

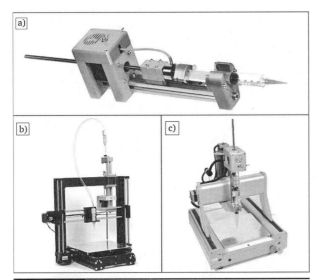

Figure 17.2 (a) The Ystruder syringe pump extruder; (b) the Ystruder mounted as an indirect drive on an open-source Prusa I3 style 3D printer using tubing; (c) the Ystruder mounted on a 43-millimeter-diameter spindle mount used for direct drive. (CC BY-4.0) https://doi.org/10.1016/j.ohx.2019.e00080

The value of FOSH in labs is not only the dollar amount saved for research but also the value of the overhead (indirect costs, which average 52 percent [in January 2013]) charged on grants to purchase the equipment. These overhead rates are primarily used to subsidize administrative salaries and building depreciation (January 2013) and have lead to a practice of rating universities by research expenditures, which perversely provides an incentive to increase these rates while depressing the use of FOSH (Pearce, 2014b). This would lead to FOSH values of more than $4.3 million and more than $2.7 million for Methods 1 and 2, respectively.

By decreasing the cost of research equipment, more resources are available to do science. For example, if a four-syringe pump is fabricated for a molecular biology lab, the savings would be enough to hire a summer student, presumably increasing the scientific discovery rate. In addition, if syringe pumps used for electrospinning novel materials represented a bottleneck to scientific discovery for a chemistry lab, being able to make between 2 and 14 open-source syringe pumps for the price of commercial alternatives would relieve that bottleneck and increase the rate of scientific discovery. Quantifying the value of an increased rate of discovery entails a specific study in each lab that would need to be done after the discoveries were made, with controls for similar labs operating with less or inferior equipment. Similarly, if because of its lower cost the pump was able to be used in the classroom or lab courses either at the university or for precollege education, the improved education that students received because of access to it would be positive, as would be the tertiary effects of their contribution to the economy. Qualitatively, however, it is clear that FOSH has scientific applications and costs less than commercial offerings has value.

The medical field is the primary market for syringe pumps, but to be used outside the developing world, the open-source pump would need to go through extensive testing and certification because of liability concerns. When the vetting and certification are complete, the effect of the decreased cost for the use of syringe pumps on the overall cost of hospital care would result in saving even more money than simply the economics of the pump. Regardless of which model is used to calculate the value of a human life, it is clear that even a fraction of a life saved would result in additional value created by FOSH development and its medical applications. It should also be pointed out that these values would tend to accrue in regions where low-cost medical equipment is sorely needed now (United Nations, 2008).

Because the designs are reusable, with solid modeling and 3D printing, designs can be expanded or joined together, rapidly increasing the rate of innovation, similar to observations seen with software (Ball, 2003). The results of this case study confirm these enormous potential values and agree with the sentiment of Nobel laureate Sir John Sulston who said that research that is open to everyone is at least nine times more valuable to society than research that is closed (Love, 2014). The internet is an example of what happens when channels of communication are left open for participation and growth, and FOSS work (Raymond, 1999) has demonstrated not only how well the internet can be used for collaborative developments, such as FOSH, but also the efficiency, profitability, and opportunities of the open-source paradigm over its proprietary counterparts.

Consumer Goods

Megasavings are not limited to things scientists want for their labs—far from it. There are now millions of free and open-source designs you can download and replicate on any number of machines such as desktop 3D printers. Consider the following three specialty consumer sporting goods that might be of interest to the athlete in all of us. First is a compound bow balancer (Figure 17.3), which helps hard-core archers to be better able to hit their targets. If you want one, it will set you back close to $300. You can 3D print it out and use about $17 worth of hardware to make your own, saving yourself well over $250, which is enough to buy a low-cost RepRap! The design has been downloaded more than 2,000 times, saving the worldwide archery community more than a cool half-million dollars already.

Figure 17.3 Compound bow balancer available on Youmagine as a design. (CC BY-SA) https://www.appropedia.org/File:Compound_Bow_Balancer_Assembled.jpg

Although you can 3D print toys, the common RepRap printers can do far more than that, even for the most challenging applications. Maybe you are more into indoor sports and like weightlifting. Once you get strong, it gets more and more irritating to load/unload the plates. To make your life a little easier, you can 3D print a dead-lift jack or an Olympic lifting jack, as seen in Figure 17.4. You can then easily strip off and reload weights onto your bar when you are by yourself, allowing you to not have to lift the bar with one hand as you struggle with removing the weight on the other side. This design has been tested with up to six plates, which is 270 pounds on one side. This should be more than enough for all but the Hulk, She-Hulk, and a few Olympic-class or professional athletes. By printing out five puzzle-like pieces with about a half a spool of filament and snapping this together, you save yourself about $32 and 7 pounds of weight from lugging a commercial metal version back and forth from the gym. Perhaps what is most impressive about this design is that it works with PLA that you can print on any low-cost RepRap. Stastica reports that there are over 38,000 gyms in the United States alone. So, even if each gym only prints one dead-lift jack, the savings for the global fitness community would be more than $1 million.

Figure 17.4 Dead-lift jack available on MyMiniFactory as a design. (CC BY-SA-NC) https://www.myminifactory.com/object/3d-print-106708

The third example is the MakeItLoud V29, which is a survival whistle (Figure 17.5). It is called the V29 because it took 29 iterations to get it just right. It is rugged (tested by throwing it against pavement, driving over it with a car, and then leaving it underwater for two hours), easy to make and carry (it prints in a single piece directly on the print bed), and most important, very loud (a raised lip holds it securely in your mouth even if you are blowing as hard as you can with no hands). The V29 gets its ear-splitting 118 decibels using two slightly different tones produced at the same time by separate chambers on either side of the device. They alternate between canceling each other out and amplifying—so you get a nice shrill whistle regardless of conditions because the harder you blow, the louder it gets. The V29 also sports a hole in the back for a key ring or lanyard so you can wear it around your neck or attach it to your zipper or key chain. Thus, the next time you go hiking, you might want to print out one of these survival whistles and keep it on your jacket. More than 425,000 other people have already downloaded it, and because survival whistles cost a few dollars on Amazon, the V29 has already saved the global community more than

Figure 17.5 MakeItLoud V29 survival whistle developed by Joe Zisa. (CC BY) https://www.thingiverse.com/thing:1179160

$1 million in distributed manufacturing savings. This is pretty good savings for just a whistle!

These, of course, are just examples, but hundreds of other open-source hardware designs have saved the global community millions of dollars—each. Many of these designs are within the reach of even novices who are willing to try to iterate a few times. (Twenty-nine is admittedly a bit outside the normal number of iterations for a device, but then again, the design is really, really good.) This is all possible because the number and variety of distributed manufacturing tools are proliferating. As more people gain access to 3D printers and CNC mills, these designs will save even more money while providing the global community with even more wealth. By each of us just giving a little of our time and effort, we can help hundreds of thousands of people. You—yes, you!—can create a million dollars of wealth as long as you share!

References

Baechler C, DeVuono M, Pearce JM. 2013. Distributed recycling of waste polymer into RepRap feedstock. *Rapid Prototyping Journal* 19(2):118–125.

Ball P. 2003. Openness makes software better sooner. *Nature* 030623-6. https://doi.org/10.1038/news030623-6

Gershenfeld N. 2005. *Fab.* Basic Books, New York.

Jan T. 2013. Harvard, MIT thwart effort to cap overhead payments. *Boston Globe*, March 17. https://www.boston.com/news/national-news/2013/03/18/harvard-mit-thwart-effort-to-cap-overhead-payments

King DL, Babasola A, Rozario J, Pearce JM. 2014. Mobile open-source solar-powered 3D printers for distributed manufacturing in off-grid communities. *Challenges in Sustainability* 2(1):18–27.

Klar V, Pearce JM, Kärki P, Kuosmanen P. 2019. Ystruder: Open source multifunction extruder with sensing and monitoring capabilities. *HardwareX* 6:e00080. https://doi.org/10.1016/j.ohx.2019.e00080

Kreiger M, Pearce JM. 2013. Environmental life cycle analysis of distributed 3D printing and conventional manufacturing of polymer products. *ACS Sustainable Chemistry and Engineering* 1(12):1511–1519.

Love J. 2014. The value of an open-source dividend. http://www.managingip.com/Blog/3390962/The-value-of-an-open-source-dividend.html

Pearce JM. 2014a. *Open-Source Lab: How to Build Your Own Hardware and Reduce Research Costs.* Elsevier, New York.

Pearce JM. 2014b. It is time to move research expenditures to the denominator in university metrics. *Academe.* https://www.aaup.org/article/it-time-move-research-expenditures-denominator-university-metrics

Pearce JM. 2015. The value of open source hardware development. *Modern Economy* 6: 1–11. https://dx.doi.org/10.4236/me.2015.61001

Pearce JM, Blair C, Laciak K, et al. 2010. 3-D printing of open source appropriate technologies for self-directed sustainable development. *Journal of Sustainable Development* 3(4):17–29.

Petersen EE, Kidd RW, Pearce JM. 2017. Impact of DIY home manufacturing with 3D printing on the toy and game market. *Technologies* 5(3):45. https://doi.org/10.3390/technologies5030045

Petersen EE, Pearce JM. 2017. Emergence of home manufacturing in the developed world: Return on investment for open-source 3D printers. *Technologies* 5(1):7. https://dx.doi.org/10.3390/technologies5010007

Raymond E. 1999. The cathedral and the bazaar. *Knowledge, Technology and Policy* 12(3): 23–49.

Wijnen B, Hunt EJ, Anzalone GC, Pearce JM. 2014. Open-source syringe pump library. *PLoS ONE* 9(9):e107216. https://doi.org/10.1371/journal.pone.0107216

Wittbrodt B, Glover A, Laureto J, et al. 2013. Life-cycle economic analysis of distributed manufacturing with open-source 3D printers. *Mechatronics* 23(6):713–726.

UN High Commissioner for Human Rights and World Health Organization. 2008. The right to health (Fact Sheet 31). United Nations, New York.

CHAPTER 18

Making the Future of Sharing

The Past: Great Minds Agree That Cooperating Builds Great Societies

No matter where or when you grew up, you are probably familiar with some form of the Golden Rule, or the ethic of reciprocity (Hertzler, 1934). It is a moral basis of human rights because there is value in having this kind of respect and caring attitude for one another. The Golden Rule and the "Law of Love" were given their classical

statements in the West by Jesus of Nazareth. Jesus said, "All things whatsoever ye would that men should do to you, do ye even so to them" and "Thou shalt love thy neighbor as thyself." Although Jesus' insights into how to live happily within a society may be the most familiar to us in the West, they are not the only, nor are they the first, statement of this moral. Religions from all over the world have some form of the Golden Rule, as seen in Table 18.1.

Table 18.1   The Golden Rule and the World's Religions

Religion	Quotes for the Ethic of Reciprocity	Source
Bahá'í faith	"And if thine eyes be turned towards justice, choose thou for thy neighbour that which thou choosest for thyself."	KALÍMÁT-I-FIRDAWSÍYYIH (Words of Paradise)
Brahmanism	"This is the sum of Dharma [duty]: Do naught unto others which would cause you pain if done to you."	Mahabharata 5:1517
Buddhism	"Hurt not others in ways that you yourself would find hurtful."	Udana-Varga 5:18
	"One should seek for others the happiness one desires for oneself."	Guatama (560–480 BC)
Christianity	"Therefore all things whatsoever ye would that men should do to you, do ye even so to them: for this is the law and the prophets."	Matthew 7:12, King James Bible
Confucianism	"Do not do to others what you do not want them to do to you."	Analects 15:23
Hinduism	"One should never do that to another which one regards as injurious to one's own self. This, in brief, is the rule of dharma. Other behavior is due to selfish desires."	Anusasana Parva, Section CXIII, Verse 8

(continued on next page)

Table 18.1   The Golden Rule and the World's Religions (*continued*)

Religion	Quotes for the Ethic of Reciprocity	Source
Inca religion	"Do not to another what you would not yourself experience."	Manco Capoc, founder of the Empire of Peru
Islam	"None of you [truly] believes until he wishes for his brother what he wishes for himself."	Number 13 of Imam Al-Nawawi's Forty Hadiths
Jainism	"In happiness and suffering, in joy and grief, we should regard all creatures as we regard our own self."	Lord Mahavira, 24th Tirthankara
Judaism	"Thou shalt love thy neighbor as thyself."	Leviticus 19:18
	"What is hateful to you, do not to your fellow man. This is the law: all the rest is commentary."	Talmud, Shabbat 31a.9:18
Native American spirituality	"All things are our relatives; what we do to everything, we do to ourselves. All is really One."	Black Elk
Roman pagan religion	"The law imprinted on the hearts of all men is to love the members of society as themselves."	Religio Romana is a modern-day neopagan religion based on the religion of ancient Rome
Shinto	"The heart of the person before you is a mirror. See there your own form."	Munetada Kurozumi
Sikhism	"Don't create enmity with anyone as God is within everyone."	Guru Arjan Devji 259
Islam Sufism	"If you haven't the will to gladden someone's heart, then at least beware lest you hurt someone's heart, for on our path, no sin exists but this."	Dr. Javad Nurbakhsh, Master of the Nimatullahi Sufi Order
Taoism	"To those who are good to me, I am good; to those who are not good to me, I am also good. Thus all get to be good."	Lao Tzu
Wicca	"An it harm no one, do what thou wilt (i.e., do what ever you will, as long as it harms nobody, including yourself)."	Wiccan Rede
Yoruba (Nigeria)	"One going to take a pointed stick to pinch a baby bird should first try it on himself to feel how it hurts."	The Yoruba people in southwestern Nigeria and surrounding areas of West Africa. Santería, Umbanda, and Candomblé evolved from the Yoruba religion.
Zoroastrianism	"Whatever is disagreeable to yourself do not do unto others."	Shayast-na-Shayast 13:29

In addition, the proverbs of primitive societies often have some form of Golden Rule. The Yorubas of West Africa say, "He who injures another injures himself"; the Moroccan tribesmen say, "What you desire for yourself you should desire for others"; and the Ba-Congo expressed the Golden Rule in a clear applied proverb: "If you see a jackal in your neighbor's garden, drive it out, one may get into yours one day, and you would like the same done for you" (Gensler, 2004). Similarly, the Upanishads of Indian Brahmanism, going back to the period around 800–600 BC, wrote: "Let no man do to another that which would be repugnant to himself; this is the sum of righteousness. A man obtains the proper rule by regarding another's case as like his own." Finally, Zoroaster (660–583 BC) believed similarly.

The Zoroastrian literature reads: "When men love and help one another to the best of their power, they derive the greatest pleasure from loving their fellow men." The major world cultures follow a form of the Golden Rule, as summarized in Table 18.2. Obviously, the gift economy galvanized by the free and open-source philosophy is directly in line with the Golden Rule. Thus, free and open-source sharing is consistent with most people's religious,

moral, and ethical beliefs. Either way (innate or learned from the great minds among us), cooperation emerges as a distinctly human endeavor. The open-source philosophy thus is well aligned with our natural tendency and our moral philosophies toward cooperation, as we discussed in Chapter 1.

Although the ancient great minds were in agreement on this principle, even some of the strongest modern minds have backed up the

Table 18.2 The Golden Rule Throughout History and Regions

Era/Region	Quotes for the Ethic of Reciprocity	Source
Ancient China	"Never impose on others what you would not choose for yourself."	Confucius (c. 500 BC)
Ancient Egypt	"That which you hate to be done to you, do not do to another."	A Late Period Hieratic Wisdom Text: P. Brooklyn 47.218.135
England	"Do not that to another which thou wouldst not have done to thyself."	Thomas Hobbs, England, seventeenth century CE
	"Don't do things you wouldn't want to have done to you. To do as you would be done by, and to love your neighbor as yourself, constitute the ideal perfection of utilitarian morality."	Humanism, British Humanist Society, John Stuart Mill, Britain, nineteenth century CE
Ancient Greece	"What you do not want to happen to you, do not do it yourself either."	Sextus the Pythagorean
	"Do not do to others that which would anger you if others did it to you."	Socrates, Greece, fifth century BCE; Plato, Greece, fourth century BCE
	"May I do to others as I would that they should do unto me. What you would avoid suffering yourself, seek not to impose on others."	Epictetus, Turkey, Rome, Greece, c. 100 CE
Ancient India	"Hence, (keeping these in mind), by self-control and by making dharma (right conduct) your main focus, treat others as you treat yourself."	Mahabharata, Wise Minister Vidura Advises the King Yuddhiśhṭhira
	"Let no man do to another that which would be repugnant to himself; this is the sum of righteousness. A man obtains the proper rule by regarding another's case as like his own."	Upanishads, the foundational document for Indian Brahmanism (c. 700 BCE)
Ancient Persia	"Whatever is disagreeable to yourself do not do unto others."	Pahlavi Texts of Zoroastrianism, Part 2 of 5: The Dadistan-i Dinik and the Epistles of Manuskihar
Ancient Rome	"Treat your inferior as you would wish your superior to treat you."	"41. Slaves," the Stoic philosophy of Seneca

ancient wisdom. There have been well-reasoned calls for greater collaboration in sciences to help move humanity forward faster. The evidence now is fairly well established that it would benefit every field to be open. This was recently explained by Nobel Prize winners Eric Maskin and James Bessen when they found that when discoveries are "sequential," patent protection discourages innovation (Bessen and Maskin, 2009). This is the exact opposite outcome hoped for by the patent system. Bessen and Maskin find that society is better off with an open approach to technical innovation. Even more remarkably, they found that even an inventor's prospective profit may actually be enhanced by competition and imitation (Bessen and Maskin, 2009). This is not overly surprising for all those familiar with the metaphor of the dwarfs standing on the shoulders of giants. This old saying means simply that you can discover more by building on previous discoveries, and it was made famous by the brilliant physicist Isaac Newton (1676) when he said, "If I have seen further, it is by standing on the shoulders of giants."

This is particularly interesting in light of what is occurring during the COVID-19 pandemic. As we discussed in Chapter 1, makers are open sourcing designs to help save people's lives and stop the spread of the virus as they radically undercut the costs of conventionally closed-source supplies and medical instruments. Again, open-source development is technically superior and seems to be the obvious choice when lives are at stake. One way to protect these makers of technologies for potential pandemics and spread open-source medical hardware even faster would be to change laws to limit liability.

An approach to do this could be to institute the Golden Rule into law by expanding Good Samaritan laws. Good Samaritan laws legally protect those who give reasonable assistance to others, whom they believe to be injured, ill,

in peril, or otherwise incapacitated (Pardun, 1997). This legal protection is intended to reduce barriers for one person to help another because of potential liability or prosecution for unintended negative consequences. These types of protections are critical to reduce barriers to companies and individuals to release all their designs, which would allow others to replicate the technology in other locations. These laws can be realized to provide as many protections as possible for patients and their caregivers from harm. Such laws could also directly help the release of documentation on known lifesaving technology. It is morally unjustifiable to refuse to simply share information that will save someone's life. One way to do this is to use an expanded version of the Good Samaritan laws. Often these laws require people to aid others who are exposed to grave physical harm if there is no risk for the rescuer.

For example, in Finland, there is an explicit "general duty to act" and "engage in rescue activities according to [one's] abilities" (Finlex, 2020). Thus, the Finnish rules include a principle of proportionality, which requires more of professionals than of laypersons. The Finnish Criminal Code Section 15 (578/1995) stipulates, "A person who knows that another is in mortal danger or serious danger to his or her health, and does not give or procure such assistance that in view of his or her options and the nature of the situation can reasonably be expected, shall be sentenced for neglect of rescue to a fine or to imprisonment for at most six months." When we apply this Finnish logic to engineers and companies that design and make medical equipment, the results are interesting. Such parties have knowledge that, if shared, would save lives, and, if not shared (or not released with a usable open-source license), there is reasonable assurance that people will die unnecessarily. Is this ethically or morally defensible? If not, it would appear that these designers and companies should be compelled to share the artificial construct of intellectual property (IP) that would save others and be held criminally liable for not doing so under such legal logic as in Finland. When analyzed in this way, the current global shortage of a wide-range of proprietary products (i.e., ventilators, N95 masks, etc.) calls into question the entire IP system. It is intuitively obvious that if proprietary designs were shared, shortages would be reduced and lives would be saved. The shortage of supplies and technologies for India during the COVID-19 pandemic analyzed in a recent study appear to support these claims (Pearce, 2020). If widespread, completely open-source products, along with their methods of production, are available, shortages do not occur.

The Future

It is always risky to predict the future. You can be virtually sure that you will be wrong. Knowing the risks, I will push ahead and look at some of the trends that I can extrapolate into the future to show some ways that the sharing ecosystem may evolve (or maybe *hopefully* evolve). We know that open-source software has become established—essentially all major companies use it for their services, and everyone who uses the internet uses it every day. To a large degree, the software world has seen the benefits of moving to free and open-source software. This is why open source is run on all supercomputers (Vaughan-Nichols, 2018), 90 percent of the cloud, 82 percent of the smartphone market, and 62 percent of the embedded systems market (Kerner, 2018). Open source appears poised to dominate the future, with over 70 percent of the Internet of Things (IoT; Hall, 2018) as well. Twenty years ago,

when open-source software was just gaining real steam, even staunch proponents such as Richard Stallman questioned the social imperative for free hardware designs (Stallman, 1999). The tools needed to easily re-create designs simply did not exist yet. Academics had barely started to consider the concept; the number of papers coming out annually on the topic was fewer than could be counted on someone's fingers. Not anymore!

Not only has the ethical authority of Stallman (2015) embraced free hardware and free hardware design, so has the academic community. Consider Figure 18.1, which shows the number of articles on open-source hardware indexed by Google Scholar each year since 2000 compared with those for open-source software. In a few short years, the concept of open-source hardware has erupted in ivory towers throughout the world. It is following the path of free and open-source software but is about 20 years behind. The curve in the figure appears roughly exponential. This indicates that not only is open-source hardware established in academia, it is a hot and growing topic that will have a substantial impact. Now, more than 1,000 articles are written on the topic every year. I should know—I am the coeditor-in-chief of *HardwareX*, a journal published by the largest scientific publisher in the world. *HardwareX* is rapidly gaining ground and is set up to be the premier instrumentation journal for scientists. Libre hardware has gone mainstream, with even the authority of the U.S. National Academy of Engineering dedicating a special issue of its magazine, *The Bridge*, to the topic in 2017.

You know from reading hundreds of examples in this book that open source is helping a lot more people than just scientists. If these trends continue, it is possible that free and open-source software will saturate all types of software and free and open-source hardware will be the

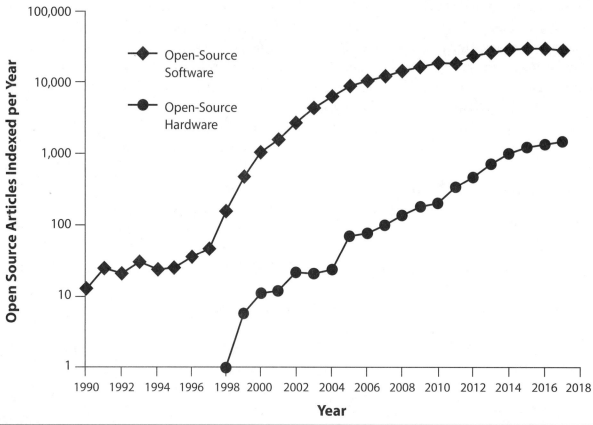

Figure 18.1   Scientific articles indexed by Google Scholar containing the phrase "open-source software" and "open-source hardware" from 1990 until 2017 shown on a logarithmic scale because of the orders of magnitude changes (Pearce, 2018).

dominant method of hardware development in every field! Although we may have a few million open-source hardware designs now, there may be billions of open-hardware designs in the future. Quite literally, every product you buy from Walmart or on Amazon now may have an open-source equivalent (or better) that you can download for free in the future. Simultaneously, we have seen distributed digital manufacturing erupt in schools, engineering firms, and manufacturing plants. If it continues to follow the same trend, all schools will have shops with 3D printers, CNC mills, PCB mills, and laser cutters. Then all libraries and community centers may look more like fablabs and makerspaces. If trends continue, eventually everyone will have access to high-quality open-source digital tools to manufacture products for themselves from these

gigantic repositories of open-source designs. As the costs have become so low (you can buy a decent RepRap 3D printer for $250 today), digital fabrication capital costs are within the price range of many middle-class families. Everyone could have the tools to make the products they want at home or in their local community. We have seen Amazon already include 3D-printing filament in its list of "AmazonBasics," which means that the company thinks it is an "everyday product." This is not happy thinking—this is what data analytics tell us from one of the largest firms in history. We can be sure that Amazon is right on this account because the company is basing this on pure data. Does Amazon know what products its customers are substituting with those they make themselves? How will businesses change? Will the last few manufacturing jobs be converted to positions like "personal fabricator"—a cross between a machinist and a personal shopper? Or will everyone prefer to do their own fabbing? Will governments step in and pay open-source designers as a means to generate wealth for their electorate? Or perhaps only a few monster companies that have learned to navigate the open-source seas will survive the relentless competition.

In the near future (or at least the not-so-distant future), it is not hard to believe that everyone will have access to open-source tools and designs to make almost all the tangible things we think of as wealth for themselves. As the pages of this book have shown, this is already happening, but there are many unanswered questions, such as

- Will they make all (or even most of) their own stuff rather than buy it?

- What if they do?

- What would society be like then?

Open-source everything would be a real option. Everything in your house, and even your house itself, could be produced by you and your family using local materials and Libre digital tools. Perhaps larger tools, such as a house printer, you would check out of the library, borrow from your local municipality, or rent from a service-based open-hardware company. Everything in your home could be personalized to a degree unthinkable now without spending enormous sums of money. If you like dragons or unicorns or a particular sports team, every product in your home could follow that theme, or you could change themes room by room. Or you could change themes every year. With digitally distributed recycling following similar trajectories, all waste could become feedstock for digital fabrication in the next generation of open-source wonderments. For example, as your daughter outgrows Hello Kitty and

becomes more interested in becoming a marine biologist, the kitty 3D prints could be ground up and reextruded into dolphins. The electronics automating the old kitty designs could be immediately reprogrammed to play calming ocean music while your daughter is studying or could be used for data loggers for buoy monitors to help your daughter collect data for her thesis.

Weird things could start to happen with humanity's interaction with the environment. Everyone's home could power itself with local renewable energy sources via microhydropower, small wind turbines, or solar photovoltaic arrays, all produced with local materials from open-source designs. In the area where I live now, the homes used to be heated with coal. No one does that now because it stinks. If you could have any technology to produce the power to run your basement fab, what would it be? Most people would probably choose something like a Tesla solar roof if cost were not an issue. Solar power is already economical, but it does take a few years to pay for itself. Costs continue to drop, and tracking solar technology out a few years, along with the rate of open-source development, it appears even the most hardcore technologies, such as the semiconductors for solar photovoltaics, could be within our open-source reach. Mammoth centralized fossil-fuel-fired power plants, already struggling to compete, would be completely overrun by green distributed generation.

Where will raw materials come from? And what about our waste? Those problems may solve themselves in a circular economy where the concept of waste is abolished. Things may even get better. Landfills could start to shrink as they begin to be mined for raw materials. This could go a couple of ways. Modern-day open-source mining companies could be set up to purchase retired landfills and then put some massive, fully automated robots into producing valuable feedstock streams. Or perhaps the future is a little more like the cyberpunk genre, where teenagers illegally break into landfills at night with their pack robots to carry off material supplies for their latest creations that needed more materials than their family's allowance.

It could truly be an entire society built with very little money but possessing enormous wealth. You can live like royalty simply because others have shared their creations with you and the rest of the world. You want to sit on a throne fancier than any of the European royalty ever even dreamed about? *Create* one! But do make sure that you *share* the design. This is how we all get to live in enormous wealth while we all *save* a boatload of money.

References

Bessen J, Maskin E. 2009. Sequential innovation, patents, and imitation. *RAND Journal of Economics* 40(4):611–635.

Finlex. 2020. Translations of Finnish acts and decrees: 39/1889. English available online at https://finlex.fi/en/laki/kaannokset/1889/en18890039

Gensler HJ, Spurgin EW, Swindal J. 2004. *Ethics: Contemporary Readings*. Psychology Press, London.

Hertzler JO. 1934. On golden rules. *International Journal of Ethics* 44(4):418–436.

Newton, Isaac. 1675. *Letter from Sir Isaac Newton to Robert Hooke*. http://digitallibrary.hsp.org/index.php/Detail/Object/Show/object_id/9285

Pardun JT. 1997. Good Samaritan Laws: A Global Perspective. *Loyola Los Angeles International Comparative Law Review* 20:591–613.

Pearce JM. 2018. Sponsored Libre research agreements to create free and open source software and hardware. *Inventions* 3(3):44. https://doi.org/10.3390/inventions3030044

Pearce JM. 2020. Distributed manufacturing of open-source medical hardware for pandemics. *Journal of Manufacturing and Materials Processing* 4(2), 49. https://doi.org/10.3390/jmmp4020049

Stallman R. 1999. On "free hardware." *Linux Today*. https://www.linuxtoday.com/infrastructure/1999062200505NWLF

Stallman R. 2005. Free hardware and free hardware designs. Gnu.org. https://www.gnu.org/philosophy/free-hardware-designs.en.html

Index